Discriminating Taste

Discriminating Taste

*How Class Anxiety Created
the American Food Revolution*

S. Margot Finn

RUTGERS UNIVERSITY PRESS
NEW BRUNSWICK, CAMDEN, AND NEWARK,
NEW JERSEY, AND LONDON

Library of Congress Cataloging-in-Publication Data
Names: Finn, S. Margot, 1981– author.
Title: Discriminating taste : how class anxiety created the American food revolution /
    S. Margot Finn.
Description: New Brunswick, New Jersey : Rutgers University Press, [2017] | Includes
    bibliographical references and index.
Identifiers: LCCN 2016025795 | ISBN 9780813576862 (hardcover : alk. paper) |
    ISBN 9780813576855 (pbk. : alk. paper) | ISBN 9780813576879 (e-book (epub)) |
    ISBN 9780813576886 (e-book (web pdf))
Subjects: LCSH: Food habits—United States—History. | Food habits—Economic
    aspects—United States. | Food consumption—United States—History. | Food
    consumption—Economic aspects—United States. | Food—Social aspects—
    United States. | Middle class—United States—Social life and customs.
Classification: LCC GT2853.U5 F565 2017 | DDC 394.1/20973—dc23
LC record available at https://lccn.loc.gov/2016025795

A British Cataloging-in-Publication record for this book is available from the British Library.

⊗ The paper used in this publication meets the requirements of the American National Standard for Information Sciences—Permanence of Paper for Printed Library Materials, ANSI Z39.48-1992.

www.rutgersuniversitypress.org

Manufactured in the United States of America

For my parents, Linda Kay and James Patrick Finn

Contents

Discriminating Taste

Introduction

DISCRIMINATING TASTE

We are living in the Age of Food.

—*Steven Poole*

Sometime around 2005, my mother called me, distressed by a conversation she'd had with one of her co-workers about wine. She had started getting into wine after I left home for college in 1999, joining a trend that was sweeping the country. Wine consumption in America had been rising gradually ever since the end of Prohibition, but from 1991 to 2005, sales jumped by 50 percent.[1] The wine and cheese reception, virtually unheard of before the 1970s, became almost as common as the cocktail hour, and words such as "Merlot" and "terroir" became part of the common vernacular. My mother's participation in this trend was relatively casual. She purchased most of her wine at the supermarkets where she did the rest of the grocery shopping, sticking mostly to a handful of national brands and varietals she knew she liked, and she rarely had more than a glass or two with dinner. However, after one particularly stressful day at work, she had consumed half a bottle by herself. When she confessed this minor indiscretion to a co-worker the next day, the woman asked what kind of wine she had been drinking. It was a white Zinfandel, one of my mother's favorites. "Oh no," her coworker sneered. "That's the hot dog of wines."

"Well," my mother said to me later, "I didn't know that. Did you know that?" I did, as she must have known I would. Although I had not yet started doing research on food in any formal capacity, I had also gotten into wine after leaving home for college. I drank keg beer at house parties and deceptively innocuous Long Island iced teas on happy hour discount, but I favored wine when it was available because it seemed to impart a kind of sophistication and worldliness I longed for. I also liked the taste and acquiring a special vocabulary to express my developing preferences. I liked learning about new grape varietals and tannins and minerality and oak. Partially due to a series of waitressing jobs at increasingly swanky restaurants and partially due to the ambient foodie-ism of my college town and then New York City, I learned to eschew not only white Zinfandel but also everything else sweet (infantile), Merlot and Chardonnay (too common), and anything with an animal on the label (pandering to the masses).[2] Still, it never would have occurred to me to say anything to my mom about her affinity for fruity pink wine. Let her drink what she likes, I figured, failing to anticipate that by doing so I might let her embarrass herself.

Even though my mother has never aspired to be a connoisseur and mostly drinks at home with no one around who might be impressed, her enjoyment of wine was still partially based on its broader cultural significance. She probably wouldn't have started drinking wine in the first place if not for the idea that it is refined or high class. I'm sure she also likes the taste, and the fact that she turned to it to unwind after a stressful day suggests that she appreciates the relaxing effects of alcohol. However, her co-worker's suggestion that white Zinfandel was déclassé ruined it for her. In that phone conversation, I think I offered some kind of lame defense about how rosés were all the rage in New York that spring and her co-worker didn't know what she was talking about, but whatever I said apparently wasn't very convincing. I don't think I've seen her drink a blush wine of any variety since.

Taste is complex—it's both deeply personal and profoundly social, rooted in our physiology but also shaped by culture and experience. Some tastes we develop as children seem to stay with us, but others change over time. Many wine drinkers find that their preferences evolve, usually moving from sweeter and lighter varietals such as Riesling and Moscato to drier whites such as Chardonnay and Sauvignon blanc and more tannic or alcoholic reds such as the "hedonistic fruit bombs" favored by wine critic Robert Parker.[3] In fact, the taste for all kinds of alcohol usually develops gradually, starting with lighter beers and wines and sweet mixed drinks. With enough exposure, many find that they enjoy hoppy India pale ales, drier cocktails, and even hard liquor neat. Even if you don't drink, you've probably experienced some kind of similar taste transformation. Most children have an aversion to all things spicy, bitter, astringent, and pungent, including hot peppers, tea, tobacco, olives, kale and Brussels sprouts, and sharp or stinky cheeses. But all of those things also offer rewards, both chemical and cultural. As people come to associate the rewards with the initially offensive stimuli, both consciously and subconsciously, their taste experience changes. The flavors become tolerable or even pleasurable.[4]

Take coffee, for example. Many people add sweetener and milk to mask coffee's natural bitterness, especially when they first start drinking it. However, as they associate the aroma and taste with the sensations induced by the caffeine and by any sugar and fat they've added and perhaps also the context where they tend to drink it—an invigorating morning ritual, a break from work, or meeting a friend at a café—most come to like coffee itself. I'd probably still find coffee repellent if I hadn't gotten a crush on a Starbucks barista in high school and been too embarrassed to order hot chocolate from him like a little girl, opting for a mocha instead. At first, the chocolate syrup and milk and whipped cream just barely made the espresso tolerable, but within a year, I was fueling late night study sessions at Denny's with endless cups of coffee, which by then I had decided I preferred black.

Not everyone follows this pattern. Even when they do, the extent of eventual acclimation varies widely. Some people actually have more taste buds than others and they tend to have a much stronger aversion to bitter tastes in general. Many people have gastronomical blind spots or sensitivities to specific flavor compounds, which is probably what makes cilantro taste like soap to roughly 15 percent of the population.[5] People also vary in how they respond to the rewards of caffeine, nicotine, alcohol, sugar, and fat. Nonetheless, there is enough commonality across individuals and cultures that the most initially offensive, hardest-to-acquire tastes—the hottest chili, the smokiest scotch, the hoppiest beer, the most pungent tobaccos, the blackest coffee—are typically associated with maturity, masculinity, and high social status. I still genuinely like the taste of black coffee, but it's hard to know how much of that is due to a reward response conditioned by the rich, sugary mochas I started with, how much to the addictive nature of caffeine, and how much to my desire for the maturity and edginess I thought it communicated.

The physiology of taste offers a partial explanation for why white Zinfandel might be considered the "hot dog" of wines. Sweet and soft, with less acid and a lower alcohol content than many wines, it's like a wine with training wheels from which many people eventually graduate, especially those who become aficionados. However, I wonder if my mother would have ever stopped enjoying white Zinfandel without her co-worker's shaming nudge or if I'd have ever acquired a taste for coffee if not for my desire to seem older than I was. Physiology alone does not explain why some tendencies become the basis for pejorative judgments about taste, and it definitely doesn't explain how the tastes of groups of people change over time. For most Americans today, the idea that wine is categorically sophisticated is a kind of common sense, almost too obvious to merit further examination. It was not always so.

From the repeal of Prohibition in 1933 through the 1960s, wine was associated primarily with poor alcoholics and immigrants. Fortified wines produced by blending brandy or grain alcohol with cheap wine or fruit juice, sugar, and artificial coloring made up the vast majority of U.S. wine sales.[6] National brands such as Thunderbird, Wild Irish Rose, and Night Train became popular during the Great Depression because of their relatively low price, high alcohol content, and sweet taste.[7] In 1940, the first year when U.S. sales reports differentiate between fortified wine and table wines, the latter constituted less than a third of the total market.[8] The sales data don't include the wine that people made at home, but that practice was never common outside of immigrant communities from Southern and Eastern European. During Prohibition, California vineyards shipped their grapes to the east, where Italian Americans bought them by the crate straight from the railroad yard. Italian Americans were also the primary consumers of the inexpensive jug wines that many of those vineyards began to produce after Prohibition, some of which are still around, such as Gallo and Carlo Rossi. Until the 1960s, jug wines were just about the only domestic table wine you could buy in America. The prevalence of cheap, low-status wines consumed primarily by low-status people was reflected in derogatory terms such as "wino" and "Dago red."[9]

In 1947, the Gallup polling agency asked Americans about their idea of "the perfect meal." Those who said they would start with an alcoholic beverage mostly said they'd prefer a "gin cocktail or plain Manhattan." The same year, approximately 60 percent of Americans told Gallup they drank alcohol at least occasionally, and most said they preferred beer or liquor.[10] By 1960, the fortified and dessert wines associated with poor alcoholics were still outselling table wines two to one. It's hard to exactly pinpoint wine's transition from lowbrow to highbrow, but based on the prevalence of table wine as a portion of overall sales, it was probably sometime in the late 1970s. Although Julia Child arrived on PBS with her enthusiasm about French wine in 1963, sales of table wine increased only modestly in the 1960s, not much more than can be accounted for by population growth. By 1970, table wine sales made up only half of the market (see Fig. 1). By 1980, they constituted three-quarters of the market, which as a whole had grown to 480 million gallons.[11] The National Minimum Drinking Age Act, passed in 1984, combined with the baby bust that followed the baby boom to briefly reduce the total population of Americans who could legally purchase alcohol, which explains the dip in overall sales in the 1980s. However, per capita wine sales continued to increase throughout the downturn.

The growing popularity of table wine in the 1970s and 1980s that Fig. 1 demonstrates was accompanied by changes in the cultural significance of

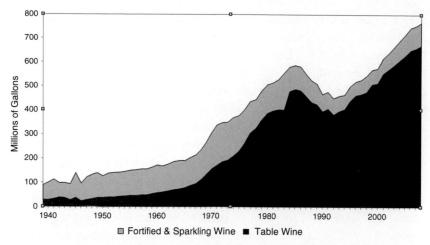

Figure 1. Wine consumption in the United States, 1940–2009. Source: Constructed from data from "Wine Consumption in the U.S.," updated April 5, 2010, *Wine Institute*, http://www.wineinstitute.org/resources/statistics/article86.

wine. Being interested in wine became a sign of refinement instead of addiction, impoverishment, or the dangerous—if sometimes alluring—sensuality and lack of self-control associated with wine-drinking immigrants.[12] In the 1990s, wine also acquired a tentative association with health, largely thanks to a *60 Minutes* episode on the so-called French paradox. It suggested that one of reasons the French have lower rates of heart disease than Americans, despite consuming more saturated fat, is that they drink more wine. In 1992, Gallup added a question to their annual survey of drinking habits about whether respondents who drank alcohol drank beer, wine, or spirits most often. That year, 47 percent said they preferred beer compared to 27 percent who preferred wine, but wine steadily ate away at its lead until 2005, when it surpassed beer as the nation's alcoholic beverage of choice. Beer recovered a slight advantage the following year, but broader trends in the beer market echo the changes in wine. Per capita consumption of the beer produced by the three breweries that dominate the U.S. market—Anheuser-Busch, Miller, and Coors—has been flat or falling since 1981.[13] The only kind of beer people are drinking more of is what is typically seen as the fancier stuff. Imported beer began to claim an increasing share of the market in the late 1990s, and since 2000, the fastest-growing segment of the beer market has been microbrews.[14]

Wine has never had just one meaning. Even in the heyday of Thunderbird and jug wines, some Americans drank imported table wine and fretted over how to use it to impress their dinner party guests. For example, a cookbook called *Dining for Moderns* published by the New York Women's Exchange in

1940 includes dinner menus with suggested wine pairings for each course, identified by the regions in France where they were produced—Champagne, Burgundy, Bordeaux—rather than the grape varietals more commonly used today. *Dining for Moderns* also criticizes hosts who served wine "simply to show off" instead of selecting the kind that would best complement the food.[15] At least for the American elite, wine may have always been associated with sophistication and thus also carried the risk of snobbery. However, there is little indication that most middle-class Americans in the 1950s and 1960s were any more interested in using wine to show off than they were in drinking Dago red. Instead, table wine was generally seen as unappealing, along with other fancy, foreign foods such as caviar and escargots. The "plain Manhattan" in the Gallup poll was preferred at least in part because of its plainness: a reliable, unfussy cocktail that would be followed by an equally plain meat-and-potatoes meal. In 1947, calling something the "hot dog" of wines probably would have been interpreted as an endorsement.

Clearly something has changed. My mother is a middle-class Midwesterner—the daughter of Japanese American sugar-beet farmers from Nebraska who now lives in suburban Chicago and works as a computer tech at an elementary school. The fact that one of her middle-class, Midwestern coworkers would even think to criticize her taste in wine (and by comparing it to a populist icon like the hot dog, no less) is the result of a comprehensive transformation in how many Americans buy, cook, eat, and above all talk about and imagine food. This sea change has been widely hailed as an American "food revolution." The signs of it are everywhere. Chefs and food writers have become celebrities with hit television shows and best-selling books. Americans collectively spend something in the range of $40–60 million every year on weight-loss programs and products, including an ever-expanding array of diet, low-fat, and low-carb foods.[16] Organic, locally grown, free-range, or otherwise environmentally friendly and ethical foods are available not only at the growing number of farmers' markets and Whole Foods Markets but also at Walmart. Options for ethnic food have expanded far beyond Italian, Mexican, and Chinese; it's no longer uncommon to find Thai, Indian, Vietnamese, Korean, and Ethiopian restaurants in Midwestern strip malls. Rice noodles, chutney, curry paste, and tahini appear on the shelves at Kroger and Safeway. Food is also the subject of a growing number of popular and acclaimed documentaries, blogs, books, college courses and even entire majors; many of the most read and most e-mailed *New York Times* articles; and a burgeoning movement for social change.

The question this book seeks to answer is: How did we get here? Perhaps the most intuitive answer is simply progress. Innovations in agriculture and food processing and cheaper, faster shipping have made a wider variety of foods available to more people than ever before. Immigrants, trade, and tourism have introduced Americans to ingredients and flavors from around the world. They are more likely to try to appreciate foods that seem foreign or ethnic thanks in part to the rise of multiculturalism and the idea that diversity is good. Additionally, a lot more information is available about how food affects our health, the environment, and the welfare of animals. If Americans didn't care as much about drinking wine, dieting, buying organic and local food, and sampling diverse ethnic cuisines thirty years ago, that may be because they just didn't know any better, care enough, or have sufficient resources before. I call this theory, which is the prevailing, commonsense explanation for the food revolution, the culinary enlightenment thesis. It attributes the changes in American food culture in the last few decades to advances in agriculture, nutritional science, ecology, and the global movement of people, goods, and ideas. According to the logic of culinary enlightenment, the food revolution is another example of new technologies, research, and wealth that enables people to improve on the deficiencies of the bad old days.

Many kinds of food have gotten cheaper, and changing immigration patterns are certainly part of the reason there are more Thai and Indian restaurants in the United States now than there were thirty years ago. But there are reasons to be suspicious of the overall narrative. After all, it's equally obvious to many people that we're living in an era of unprecedented culinary backwardness and on an overall downward trajectory. We are constantly told by public health authorities and the mass media that Americans hardly have anything worthy of the name *cuisine* and instead prefer nutritionally empty junk food with little regard for the supposedly dire long-term health effects and consequences for the environment of consuming such food. There's a popular alternative to the prevailing story of progress that I call the culinary decline thesis. Instead of seeing new technologies, information, and trade as the source of better-tasting, healthier, safer, fresher, more varied, or more environmentally friendly foods, the decline thesis argues that industrialization and capitalism have distanced people from the agricultural origins of their food, polluted the environment and the food supply with toxic chemicals, and eroded traditional social institutions such as the family farm and the family meal. As a result, most Americans today don't know what real food tastes like or how to prepare it.[17] They are thus reduced to subsisting on highly processed convenience foods that make them fat and sick. Instead of putting faith in the inexorable forward march of progress, proponents of the decline thesis often promote a return to the good old days. This romantic yearning for the past is captured in the idea that people need to get back to the land, to get back in the kitchen, or—as Michael Pollan advises—to avoid "anything your great-grandmother wouldn't recognize as food."[18]

There are probably some kernels of truth in the decline thesis. There are fewer Americans employed directly in agriculture today than there were fifty or a hundred years ago. The consolidation of food processing has made it theoretically possible for a single source of contamination to sicken hundreds or even thousands of people across the country. And the American population as a whole has gotten slightly fatter—about ten to twenty pounds, on average, between 1980 and 2000, though there has been no statistically significant increase in the last fifteen years.[19] However, the significance of these changes may not be as dire nor the past as rosy as the decline thesis suggests.

Small, diverse family farms such as those idealized in films such as *Food, Inc.* were never the dominant form of agricultural production in the United States, particularly in the South. Even Thomas Jefferson, perhaps the most famous advocate of the self-sufficient yeoman farmer, owned a large plantation that depended on slave labor.[20] The overall rate of food-borne illness has dropped by nearly 25 percent since the Centers for Disease Control began tracking laboratory-confirmed infections caused by the six most common pathogens in the late 1990s.[21] Despite continued industrial consolidation in that period, the only food-borne illness that is more common now than two decades ago is caused by the *Vibrio* bacteria, which is contracted primarily by eating raw oysters and may be increasing due to rising ocean temperatures.[22] Despite decades of public health warnings that fatness is going to cut our lives short, average life expectancy has continued to increase throughout the so-called obesity epidemic and many measures of health supposedly related to weight have improved.[23] Furthermore, there's little evidence that eating real food and engaging in vigorous daily exercise would make everyone thin. The Old Order Amish spend an average of ten to twelve hours a day performing tough physical labor and eat largely the way their great-great-grandmothers did, but they're just as fat, on average, as the rest of the American population. Amish women over the age of forty are actually slightly fatter.[24]

Although the culinary enlightenment and decline stories may seem to be diametrically opposed, they actually have much in common. Both portray the industrial food system as the source of fundamentally bad food— homogeneous, highly processed, artificial, fattening, nutritionally and environmentally toxic crap. Even the fact that many people seem to like how this "junk" food tastes is generally portrayed as a bad thing, although the reasons why can be contradictory. Some critics of industrial food argue that it is ingeniously engineered to be so intensely flavorful that it ruins people's ability to appreciate simpler, subtler, more authentic flavors. Others claim that the inferior quality of industrially grown and processed ingredients renders the food flavorless compared to heirloom crops and homemade products. Whether it is deemed excessively flavorful or flavorless, industrially processed food is consistently portrayed as inferior to real food.

The enlightenment and decline stories also share the same basic criteria for what made the foods we supposedly used to eat or the ascendant trends of the food revolution better. In addition to being more authentic, they're supposed to be fresh, whole, natural, local, organic, sustainable, traditional, diverse, artisanal, and healthy. It's not always clear where exactly the line between bad, processed junk and real food ought to be drawn. For example, homemade bread made with packaged yeast and white flour or Lay's Classic potato chips made with only potatoes, vegetable oil, and salt might reasonably belong to either category. But those are minor definitional quibbles. In terms of their overall aesthetics and ethics—what kinds of food are good, what kinds of food are bad—the culinary enlightenment thesis and the culinary decline thesis are aligned. The real difference between the stories is what kind of food they think is winning. Are industrialization, capitalism, and mass media working to improve or degrade America's food culture as a whole? Are we a fast food nation or nation of foodies? My goal is not to settle that debate. Instead, this book seeks to account for the rise of the particular set of beliefs about what kinds of food are good and bad that both the enlightenment and decline narratives share.

Those beliefs are part of the ideology of the food revolution, a comprehensive way of thinking about food that's become the prevailing common sense in America. How did so many people come to see things such as wine, kale, and Thai food as desirable? Progress isn't necessarily the most convincing answer.

Instead, the ideology of the food revolution may have been embraced because it appealed to many Americans whose income and wealth had begun to stagnate or decline in comparison to the soaring fortunes of the super-rich. From 1880 to 1920, the conspicuous consumption of gourmet and ethnic foods, dieting, and social reform efforts such as the pure food and drug movement offered the emerging professional middle class a way to distinguish itself from the lower classes and claim moral authority over the industrialists and corporations that were eclipsing them in wealth and power. It seems at the very least possible that the correspondence between income inequality and the popularity of four central tenets of enlightened eating— both in the Gilded Age and since the late 1970s—is not coincidental.

From 1930 to 1980, the middle classes made significant gains in wealth and cultural power and their status anxieties changed. The rewards of culinary distinction faded and concerns about the softening and corrupting influence of affluence increased. Instead of dinner parties with multiple courses paired with wine, the plain meat-and-potatoes meal became the culinary emblem of the good life.[25] Instead of seeking the gourmet or exotic, most Americans began to regard culinary frills and foreignness with suspicion or disgust. Natural foods became a fringe, niche market associated with long-haired, unpatriotic "nuts" and "flakes,"[26] and even weight-loss dieting declined in popularity as a more voluptuous body ideal returned. It's not that the middle class stopped using food to perform a desirable class identity. Instead, there were shifts in what kind of class identity was seen as desirable and how people used food to communicate that. The symbolic vocabulary of class and the nature of class aspiration changes in response to changes in class structure.

The contemporary food revolution started around the same time as what the economist Paul Krugman calls the great divergence in wealth and income distribution, a reversal of the great compression in incomes that began with the Great Depression. Beginning in the late 1970s, the middle class began to stagnate as the richest 1 percent started getting much richer. Shortly thereafter, gourmet and natural foods began to gain traction, first among young urban professionals on the coasts and more gradually in the mainstream, middle-class American heartland. Diet food products began to appear on regular supermarket shelves and a fitness craze swept the nation as a slimmer beauty ideal reminiscent of the turn-of-the-century Gibson girl and flapper became popular again.[27] Other trends that had also been marginal since the Great Depression, from drinking fine wine to eating whole grains, entered the mainstream. A preference for foods constructed as elegant, virtuous, and exotic began displacing the mid-century preference for the familiar, plain, and ample.

I can't prove that people started eating "better" again primarily because of class anxieties, but it's the explanation that seems to best fit the evidence, not just in terms of the history of the central ideals of the food revolution but also in terms of the way people portray and talk about those ideals today. The resurgence of interest in enlightened eating has spurred a great proliferation of mass media about food. Restaurant reviews and recipes in the *New York Times* have grown from a minor feature to an entire food section, rivaling the paper's coverage of sports and the arts. Food television, which used to consist of a handful of didactic cooking shows on PBS such as *The French Chef* and *The Frugal Gourmet*, has diversified to include restaurant make-over shows, cooking competitions, and globe-trotting documentaries, all broadcast around the clock on multiple cable networks devoted entirely to the subject and on major network and cable channels. The most popular food blogs and recipe sites attract millions of visitors, and some food writers who started online such as Deb Perlman of Smitten Kitchen and Kenji Alt-Lopez of Serious Eats have now published best-selling cookbooks, one of the lone sectors that is still growing as book sales in general decline.[28]

These diverse kinds of food media reveal patterns in the way some foods have been constructed as good and others bad. Three kinds of stories in particular stand out for their ability to make food a viable arena of class aspiration: 1) making taste and thinness into meritocracies; 2) denying that trying to eat well is about status; and 3) portraying value-added consumption as virtuous rather than profligate. Middle-class Americans facing stagnating wealth and frustrated upward mobility have embraced these stories because they offer another way to reach for the good life. Eating "better" offers com-pensations for the lack of material gains in the last few decades. I call this phenomenon aspirational eating, a process in which people use their literal tastes—the kinds of foods they eat and the way they use and talk about food—to perform and embody a desirable class identity and distinguish themselves from the masses.

The word "ideology" is sometimes used to refer to explicit, conscious belief systems, usually in the realm of politics. It is often used to distinguish between rational or moderate beliefs and those seen as irrational or extreme, which are often attributed to an "ism" such as fascism or neoliberalism. That's not what I mean when I talk about the ideology of the food revolution. Instead, I'm using the word to refer to particular understandings of the world that are taken for granted as totally obvious at a particular moment in time.[29] Unlike most isms, which people often embrace or reject deliberately, ideolo-gies in this sense of the word are generally implicit. They shape how people perceive reality and what kinds of stories they tell themselves about who they are and why things happen, but people usually aren't conscious of them. They surround us invisibly like the air we breathe. Consider the beliefs that wine is sophisticated and that sophistication in food and drink is a good thing. Those ideas are so commonly taken for granted right now that most people aren't even aware many people could (and until recently did) see wine as categori-cally low class and fancy foods as undesirable.

Ideologies can't be directly observed or measured,[30] but they can be inter-preted from the things people say, do, and create. Mass media are particularly rich sources of clues about what kinds of beliefs are winning the struggle over meaning in a particular time and place because they tend to reflect the popular understanding of the world in order to appeal to a broad audience. Mass media also shape how people see the world. In other words, mass media and dominant ideologies are mutually constitutive; the influence runs in both directions. But the relationship isn't always simple or direct. The world view represented in a particular news story or television show is inevitably partial, and any given film or blog entry may contain contradictory or unstable ideological elements. Even when the ideology represented in a mass media text is straightforward, audiences don't necessarily embrace the world view they encounter in mass media whole cloth or believe everything they see. People sometimes appreciate things in an arch or ironic fashion or interpret something as humorous that was meant to be serious.

Whenever possible, I have paired examples from mass media texts with evidence of their actual audience reception, often from the comment threads on articles, blogs, fan message boards, and review sites. The advent of forums such as these provides a large and ever-growing source of information about how people are interpreting mass media, but as a source of reception data it is not without problems. For one, the participants in online forums are probably not representative of the entire audience. The people who comment on articles or review movies on sites such as the Internet Movie Database (IMDb) are generally those who have stronger feelings than most; they are the ones who really like or really hate whatever they're writing about. They are also likely to be somewhat younger, wealthier, and better educated than the average viewer. Nonetheless, especially considered in aggregate, they can provide a sense of how people are making sense of media. When multiple contributors to several different forums express a similar interpretation of a given text, it's probable that other viewers have a similar take. One advantage online forums have over the kind of experiments and surveys that prompt people for reactions to media is that there are no measurement or observer effects. Additionally, comments in online forums capture audience reactions to media however they're encountering it in their own lives with any effects of the influence of their friends and family and the broader cultural context intact. To complement the reception data from online forums, I also use Nielsen ratings, opinion surveys, and sales data in order to better understand how people are using and reacting to mass media and the ideologies represented in it.[31]

Although I focus on cultural products mostly created by and aimed at Americans, the food revolution and aspirational eating are not exclusively American phenomena. The epigraph declaring this to be the "age of food" comes from *You Aren't What You Eat: Fed Up with Gastroculture* by British writer and *Guardian* columnist Steven Poole.[32] Many of the targets of Poole's ire, such as Jamie Oliver, are celebrities throughout the Anglophone world. *MasterChef Australia*, the antipodean version of a competitive cooking reality show that originated in Britain, has been one of the country's most watched

programs since its 2009 debut.[33] The vibrant restaurant scene, farmers' markets, and food activism in Toronto are similar to what you see in American
cities such as Portland, San Francisco, and New York.[34] Some aspects of the
food revolution may even be global. The Slow Food movement, founded in
Italy in 1986, now has chapters in 150 countries around the world. Japan is
home to more restaurants with three Michelin stars than any other country,
and trend-setting bars in the Ghanaian capital of Accra feature local ingredients with a cosmopolitan twist, such as *caipirihnas* made with local palm
spirit and locally caught salmon garnished with avocado velouté.[35]

If my theory about the relationship between class structure and food
culture is correct, this makes sense, as the economic trends that have shaped
American class anxieties since the late 1970s are not restricted to the United
States. Technological changes, globalization, and neoliberal economic policies have affected many countries in similar ways. The Reagan revolution was
paralleled by Thatcherism in Britain and economic rationalism in Australia,
all of which idealized free markets, ushered in a relaxed regulatory environment, privatized state industries, increased restraints on unions, cut taxes and
social support programs, and reduced barriers to trade. Since the mid-1980s,
income inequality has increased in seventeen of the twenty-two member
nations of the Organisation for Economic Co-operation and Development.[36]
Even in historically egalitarian countries such as Denmark, Germany, and
Sweden, the divide between the rich and poor has widened. Some of the
global manifestations of the food revolution may be the result of American
influence, but some are a response to other changes in politics, economics,
and culture.

The United States itself is vast and heterogeneous, so the flip side of the
fact that many American cultural trends are not unique to America is that
even the most popular ones are irrelevant to many Americans. It's hardly a
secret that drinking wine that costs $30 or more per bottle and that veganism,
passionate locavorism, and culinary tourism are all more prevalent in cities,
on the coasts, and among the college-educated, affluent, and liberal. However,
even if the population most active in the trends that make up the food revolution is actually a small and unrepresentative minority, it is a minority that
has a disproportionate influence on mass media and dominant cultural ideas
about taste and morality. This is largely a story about the mainstream popular
discourses about food that typically reflect the desires and anxieties of the
dominant social classes.

The core constituency of the food revolution is the group sometimes categorized as the professional middle class or the professional and managerial
class. As with all contemporary social classes, the exact dimensions of the

professional and managerial class are unclear. Sociologists such as William Thompson and Joseph Hickey estimate that it consists of roughly 15 percent of the American population that is largely in the top quintile of households by wealth and income, but not the top 2–5 percent.[37] However, social class is not defined exclusively by income and wealth and is better understood as a cultural distinction than an income bracket. Members of the same social class tend to share a broad set of beliefs about how one should dress, speak, eat, decorate, work, play, worship, and so on. For example, academics (and particularly the growing number of contingent faculty) and journalists often earn far less than many skilled tradespeople and yet are more likely to share the tastes, values, and habits of the professional and managerial class. Although numerically a minority, this class is special because, as Barbara Ehrenreich writes, it "plays an overweening role in defining 'America': its moods, political direction, and moral tone."[38]

However, although the professional and managerial may be the primary class that is participating in and proselytizing for the food revolution, the shift in American attitudes toward food certainly isn't limited to them. The widespread popularity of television shows such as *The Biggest Loser*, films such as *Sideways*, books such as *The Omnivore's Dilemma*, and aspirational foods themselves, from wine to pad thai, attest to its reach.

In that sense, the food revolution has been a great success. Organic farming and farmers' markets may represent a small part of our national agriculture and grocery systems, and many more Americans qualify for federal nutrition assistance than can ever afford to eat at restaurants such as Chez Panisse or Topolobampo. Nonetheless, the ideology of the food revolution provides the dominant framework for today's popular discourse about food. It has become a form of common knowledge that gourmet food tastes better, thinness is healthier, organic and locally grown foods are more sustainable and ethical, and the little hole in the wall run by immigrants is more authentic than a chain restaurant. When I started researching food in graduate school, I more or less believed all of those things too.

This book grew out of an attempt to explain how so many people—including me—came to know those things despite the fact that in some pretty fundamental ways they're basically wrong. Even wine experts can't reliably distinguish between expensive and cheap wines or even between white and red when they're disguised. Weight-loss diets almost always fail to make people thinner in the long term. In fact, they fail so reliably that there's no conclusive evidence that long-term weight loss would actually make the average fat person healthier. Some of the locally grown food at farmers' markets actually travels more miles per unit than supermarket produce from across the country, even before taking into account any extra car trips on the consumer's end to get it. Whether fertilizers and pesticides are organic or not has little to do with their actual effects on the environment or human health. And it's hard to even assess the accuracy of claims about "authenticity" in food because the term itself is such a moving target.

None of those issues discount the possibility that people might cultivate new taste preferences or a better understanding about nutrition or ecology that might guide them to make food choices with positive effects on their lives or communities. One of the core convictions of this book is that trying to eat well, whether that means having roast beef with instant mashed potatoes or curry with locally grown organic vegetables, can be a profound source of pleasure and meaning. However, one of the consequences of the food revolution is that some ways of trying to eat well have been granted more social legitimacy than others, and not because they are actually better. I suspect that many people would be a little less eager to spend time and money trying to live up to the ideals of the food revolution if they knew how badly most idealized foods miss their intended mark and how contradictory the ideals themselves are.

My theory that the ideals of the food revolution gained traction due to class anxiety builds on a growing body of work in the fields of food studies and fat studies that is critical of dominant beliefs about food. It also seeks to explore some of the cultural consequences of the dramatic shifts in income and wealth inequality documented by economists such as Thomas Piketty and Emanuel Saez. Even scholars who don't specialize in food have sought to explain the growing public interest in eating better. For example, sociologist Barry Glassner's 2007 book *The Gospel of Food* poses the question: "Why do we deify some meals and some foods, and demonize others?" Like me, he concludes that many of the claims people make about how we should eat are nonsense, particularly when it comes to nutrition, in part because food marketers and advocacy groups often hijack nutritional science to further their own agendas. However, he attributes the popular embrace of these nonsense claims to a loss of faith in medicine and a broader sense of unease about "the bad order of society."[39]

The idea that our food-obsessed age has more to do with class anxiety is closer to Charlotte Biltekoff's argument in *Eating Right in America* that dietary reform movements offer the middle class a way to affirm its status.[40] Helen Zoe Veit's *Modern Food, Moral Food* also argues that since at least the early twentieth century, middle-class Americans have "used nutrition knowledge as a badge of their own education, modernity, and privilege."[41] Both Biltekoff and Veit suggest that the historical nutritional discourses they examine have implications for contemporary ideologies and reform efforts. This book explores some of those implications by looking more closely at how class operates in the food revolution. It also broadens the scope beyond nutrition and dietary reform to include gourmet food, sustainable diets, and ethnic cuisines. Much like eating right in terms of health, the professional and managerial class has used its construction of superior taste, ethical consumption, and cosmopolitan dining to affirm its status and perform its knowledge and cultural power.

Some food scholars have come to different conclusions about how social class has influenced historical trends. In *Turning the Tables*, Andrew P. Haley attributes the rise of cosmopolitan dining in the Gilded Age to the frustration of the rising middle class with their exclusion from upper-class culture. According to him, the middle class was angry and alienated by "the aristocracy's unbridled enthusiasm for all things European" and embraced ethnic food to "undermine the hegemony of aristocratic tradition."[42] Focusing on a particular aspect of food culture such as nutrition or restaurants makes it possible to do a far more granular analysis of those trends. However, one of the advantages of casting a wider net is the ability to evaluate whether potential explanations about the causes of such trends make sense in the broader cultural context. Haley's argument that the middle class pursued cosmopolitanism in an attempt to democratize restaurant dining and counter aristocratic influence is difficult to reconcile with the simultaneous Francophilia and the middle class's emulation of elite practices such as *service à la Russe* that's evident from cookbooks and women's magazines aimed at middle-class readers.

This book's attention to the longer history of culinary ideals such as sophistication and cosmopolitanism also leads me to several slightly different conclusions than several other recent books have reached on the relationship between food and social class, such as Josée Johnston and Shyon Baumann's *Foodies* and Peter Naccarato and Kathleen LeBesco's *Culinary Capital*. I agree wholeheartedly with Johnston and Baumann's argument in *Foodies* that authenticity and exoticism have emerged as key indicators of value in what they call the gourmet foodscape and I share their concern about the ways food perpetuates social hierarchies. However, they attribute the rise of contemporary foodie culture primarily to concerns about the industrial food system, globalization, and the rise of new media. The fact that similar trends emerged in the Gilded Age suggests that the mainstream preoccupation with food both then and now has more to do with status than with any of those factors. Additionally, while they suggest that the gourmet foodscape has a genuinely egalitarian and progressive pole that exists in tension with its "bourgeois piggery," I argue that the relationship between those aspects of food culture has always been more complementary than oppositional. Like the practice of slumming, which also dates back to the Gilded Age, selectively appreciating low-status foods mostly serves to create the illusion of inclusivity without really elevating the stuff that is seen as what poor people eat as a matter of course.[43] Similarly, taking a longer historical view of mainstream beliefs about food makes me somewhat less optimistic about the potential resistance to culinary norms that Naccarato and LeBesco see in competitive eating events and "junk food" blogs and websites.[44]

My conclusions about what the history of food beliefs means for contemporary food reform efforts are similar to the ones Aaron Bobrow-Strain reaches in *White Bread*, which explores the fraught history of America's relationship to industrial food through changing attitudes toward factory-made bread. According to Bobrow-Strain, "dreams of good food play a unique role in the creation of social distinctions" and "real change will happen only when well-meaning folks learn to think beyond 'good food' and 'bad food,' and the hierarchies of social difference that have long haunted these distinctions."[45] The use of food as a compensatory arena of aspiration may have answered real cultural needs created by the long decline in the middle-class share of American wealth. However, I suspect that it also exacerbates growing economic inequality by stigmatizing the foods associated with the working class and reinforcing the idea that elite consumption is more virtuous and enlightened.

Knowing a little more about how the ideology of the food revolution became dominant might help people make better choices about how to use food to pursue their personal goals or simply make them less anxious about liking things such as white Zinfandel. I hope it will also make people more cautious about passing judgment on other people based on how they eat. Food has always been part of the performance of social status, but in the last three decades in the United States it has taken on an outsized role. As long as the consumption of gourmet, slimming, natural, and ethnic foods serves as a marker of belonging to the social elite, it will continue to reinforce our widening class divisions instead of forming the basis of any real revolution.

CHAPTER 1

Incompatible Standards

THE FOUR IDEALS OF THE
FOOD REVOLUTION

You won't see Broiled Pork Chops on a menu. Instead, it's Organic Heritage Pennsylvania Center-Cut Pork Loin Chop Broiled a la Plancha with a Soubise of Toy Box Tomatoes, Hydroponic Watercress, Micro Arugula, accompanied by a Nougatine of Spring Onions, garnished with a Daikon Escabèche, topped with Prune Essence and Juniper Foam.
Want fries with that?

—*Kate Workman*

In 2009, a *New York Times* headline asked: "Is a Food Revolution Now in Season?" The answer implied by its litany of examples was yes: the Obamas had planted a garden on the White House lawn, Walmarts across the country were selling organic yogurt, and celebrities from Oprah to Gwyneth Paltrow had become advocates of food variously described as healthy, local, organic, sustainable, authentic, clean, slow, real, or simply good. However, it was somewhat unclear whether everyone involved in this so-called revolution had embraced the same cause. The movement seemed to include everyone from nutritionists seeking to improve public health to chefs who think local ingredients are more delicious to environmentalists hoping to protect fragile ecosystems to consumers trying to support their local economy. However, these varied groups and aims are often seen as part of a broader gestalt.

Shopping at farmers' markets seems to go hand in hand with eating humanely raised meat, avoiding high-fructose corn syrup, and knowing how to appreciate exotic superfoods such as quinoa and acai berries. Many of these trends promise to address multiple concerns simultaneously. Produce grown on small, local farms is supposed to be better tasting, healthier, and more environmentally friendly than fruits and vegetables transported from distant industrial-scale farms. The same goes for organic (versus conventional) crops, grass-fed (versus corn-fed) beef, free-range (versus battery cage) eggs, and cooking from-scratch with whole foods (versus processed junk and fast food). These are all portrayed as the alternatives to the standard American diet and promoted as a way to counter problems ranging from the obesity epidemic to climate change to the supposed debasement of American cuisine.

Kate Workman's parody of menus that detail every aspect of a dish's pedigree and preparation captures the broader culinary zeitgeist of the food revolution and its simultaneous aspiration to achieve aesthetic refinement, nutritional superiority, ethical and ecological responsibility, and cosmopolitan chic. In fact, any snippet could probably stand in for the whole; the local, heritage pork chop evokes not just the concern with natural or sustainable food production but also beliefs about taste, health, and authenticity. Similarly, the presence of essences, foams, or fancy terms such as *soubise* and *escabèche* on a menu are not only signs of gourmet cooking but also go hand in hand with the use of ingredients typically supposed to be healthier, more sustainable, and novel or exotic. The gestalt you can invoke with a single word, such as "artisanal" or "kale," often feels like a coherent aesthetics and ethics of eating. However, upon closer examination, the ideals of the food revolution aren't always so compatible.

Quinoa, for example, has become popular in the last decade with the professional middle class because it's supposedly healthy and has some foreign cachet. It frequently appears on menus that advertise their commitment to locally grown foods, despite the fact that virtually all quinoa consumed in the United States comes from Bolivia, where consumption of the former staple has fallen in the last decade due to increased prices driven by international demand.[1] The local and the exotic are frequently at odds. Similarly, although it might be possible to cook a meal that's both gourmet and still sounds healthy to most people—Workman's daikon escabeche (essentially pickled radish) sounds pretty virtuous—frequently the former involves the use of supposedly fattening ingredients such as meat and cream and a progression of courses that cumulatively contain far more calories than an average fast food meal.[2] Although some people might find steamed kale with brown rice delicious, that wouldn't fit many peoples' definitions of gourmet and it isn't authentic in any historical, regional, or ethnic culinary tradition. The foods celebrated by the food revolution often fail to live up to their billing. In many cases, food that's supposed to be better tasting, healthier, or more ethical isn't really any of those things. Theoretically, it would be possible for the ideology of the food revolution to be flawed but still coherent. Even if all these trendy new foods were everything they've been made out to be, their ascendance still wouldn't be evidence of progress because the ideals they represent are contradictory in some fundamental ways.

The fault lines between the ideals come into especially stark relief when someone tries to live up to all them at the same time—or, as was the case for Barack Obama during his first campaign for the presidency, when someone has multiple ideals projected onto them. The conflicts that emerged in various efforts to portray Obama as either a model eater or a man of the people and to divine his personality and politics from his food choices expose many of the tensions and fissures in the presumed gestalt of the food revolution. Instead of representing a coherent, unified ideology about what makes food better that became dominant as people achieved culinary enlightenment, it turns out that the popular discourse about food consists of conflicting and competing ideologies about what makes food "better" that even dedicated foodies often disagree about.

Those competing ideologies cluster around four central ideals: sophistication, thinness, purity, and cosmopolitanism. In the popular discourse about Obama and food, these ideals took the form of concerns about 1) his affinity for gourmet or upscale foods; 2) his body size and its relationship to his eating habits and health; 3) his knowledge of natural foods; and 4) his appreciation for diverse regional and ethnic cuisines. The four central ideals aren't mutually exclusive and indeed often get invoked simultaneously. For example, Obama's thinness was often attributed to his preference for upscale or natural products. However, it was also associated with foods like the MET-Rx brand protein bars Obama reportedly favored on the campaign trail as occasional meal replacements, which don't quite pass muster as either gourmet or natural. The popular discourse about Obama reveals how each of the four ideals structures the way people talk about food today and exposes fault lines in the apparent gestalt of the food revolution.

The lack of ideological coherence among the four central ideals of the food revolution fundamentally undermines the theory of culinary enlightenment. If the revolution were really the result of progress in terms of the information and foods available, its central ideals ought to be compatible. Instead, these conflicts challenge the very idea that these foods are actually better. That doesn't mean that the growing interest in eating better isn't part of a broader gestalt; it just means that it isn't a gestalt unified by a coherent ideology. The popular sense of culinary zeitgeist is still significant, and it is worth trying to understand how an array of trends with no ideological coherence can be invoked by a single ingredient such as kale. The popular discourse about

Obama and food suggests that what these four ideals have in common is not any logical compatibility but instead their association with a particular demographic—the professional middle class, or even more narrowly, the left-leaning part of it sometimes referred to as the liberal elite.

Obama's affiliation with the more liberal of the two major political parties and his embodiment of upward mobility are probably what encouraged people to see him as a model eater, even more than previous presidents or the other candidates he ran against. However, the expectation of many food enthusiasts and activists that Obama would share their tastes and champion their causes was frustrated. In part, that was due to the inherent conflicts between the ideals, but it was exacerbated by Obama's tendency to defy the "enlightened" choices associated with the liberal elite, often in favor of ones more likely to be seen as populist. Both the attempt to make Obama a model eater and his resistance to that portrayal suggest that what really unifies the four ideals of the food revolution is what they communicate about class. The ideals seem like a gestalt because they all do the same kind of cultural work: gourmet, healthy, natural, and ethnic foods all communicate a distance from mainstream tastes, providing a kind of symbolic image boost. It makes sense that this ability to distinguish yourself through your tastes would have had growing appeal since the 1980s as a period of widely shared economic growth and prosperity gave way to income stagnation for the masses and increasing wealth disparities. At least that seems more plausible than the idea that a collective enlightenment prompted the embrace of both skim milk and triple-crème brie, of local produce and imported beer and wine, of Jenny Craig prepared meals and all-you-can eat Indian buffets. Both the rise of the four ideals of the food revolution and the ambivalence about them are explained by the use of food as a status symbol.

THE FRUSTRATED CAMPAIGN TO MAKE OBAMA AN IDEAL EATER

At 8:15 P.M. Pacific time on November 4, 2008, just moments after CNN officially called the presidential election in favor of Barack Obama, Eddie Gehman Kohan launched a blog named *Obama Foodorama*. She calls the site "the digital archive of record about the Obama Administration food and nutrition."[3] In March 2009, Kohan claimed to be getting far more page views than several other popular policy-oriented food blogs such as *Civil Eats* and *The Ethicurean*.[4] Her blog also attracted the attention of mainstream media outlets, including the *New York Times*, the *Washington Post*, the Associated Press, *The Today Show*, and MSNBC. Kohan told the Associated Press that on the day after the Obamas' May 2009 date night in Manhattan, when they dined at a trendy farm-to-table restaurant in the West Village named Blue Hill, *Obama Foodorama* got "millions of hits."[5]

Presidential eating habits have occasionally attracted popular attention in the past. The first *White House Cookbook*, published in 1887, was commonly given as a gift to new brides at the turn of the century and remained in print until 1996.[6] Jefferson was known for his predilection for French food, and the Kennedys made headlines when they hired the French chef René Verdon to cook for them. The first President Bush's hatred of broccoli became infamous after he banned it from Air Force One, saying although his mother had made him eat it as a child, now "I'm President of the United States, and I'm not going to eat any more broccoli!"[7] However, the intensity of popular interest in presidential food preferences catalyzed by Obama's nomination and election in 2008 seems truly unprecedented. Throughout the campaign, his eating habits were scrutinized for evidence of a food reform agenda. On the day of his inauguration, the most visited page of the inaugural website was the lunch menu.[8] When the *New York Times* asked her to explain the interest in Obama and food that inspired her to start her blog, Kohan said: "He is the first president who might actually have eaten organic food, or at least he eats at great restaurants."[9] Her reasoning is emblematic of the imagined gestalt of the food revolution: she doesn't really know if Obama eats organic food but assumes as much because she knows he eats at "great restaurants." Although Obama himself never really expressed any special interest in food and has yet to take any notable stands on food policy, even as his presidency comes to an end, Kohan projected her own excitement about the burgeoning food revolution onto him.

The size of her audience suggests that Kohan wasn't alone. News outlets generally portrayed Kohan's blog as just another example of the hysterical celebrity culture that developed around Obama. However, her site offers more than just breathless teen magazine–style reporting on the president's likes and dislikes. The entries range from commentary on where and what members of the First Family have been seen eating to detailed biographies of Obama's appointees to agencies such as the FDA and the USDA and thorough analysis of all manner of food-related legislation. The breadth of her focus also testifies to the novel character of the popular interest in Obama's eating habits. Particularly during his first presidential campaign and initial years in the White House, there was a lot of excitement and curiosity about Obama, not just as an eater but also as a potential reformer of the U.S. food system.

As his time in office continued, much of the initial excitement waned. Michelle Obama embarked on a campaign against childhood obesity called "Let's Move!" and planted a garden on the White House lawn, but the president himself never embraced the "Farmer-in-Chief" mantle that Michael Pollan urged him to adopt in a *New York Times* op-ed that topped the lists of most read and most e-mailed stories for weeks Instead, he has remained a culinary cypher, occasionally engaging in the latest culinary trends but other times seeming to deliberately rebel against the food revolution. In the latter case, he may have been making calculated efforts to appeal to a broader constituency, but a closer look at the hopes that were projected onto him suggest that his betrayal of the food revolution faithful may have been inevitable. There was simply no way for him to live up to all of the ideals of the food revolution, even if he had wanted to.

The Ideal of Sophistication

The popular association of Obama with fancy foods may have started with a quote that had nothing to do with his own eating habits. In 2007, his campaign organized a "Rural Issues Forum" at the Van Fossen farm, which is located about thirty minutes outside Des Moines, Iowa. Obama spent about an hour there speaking with a group of local residents about their concerns, and one of the recurring themes in the conversation was farming. After noting that the prices most farmers receive for their crops had not increased despite rising food prices, Obama said: "Anyone gone into Whole Foods lately and see what they charge for arugula? I mean, they're charging a lot of money for that stuff." The *New York Times* blog *The Caucus* reported that the line "landed a little flat," and suggested that may have been due to the fact that the closest Whole Foods Markets were across state lines in Minneapolis and Kansas City.[10] Arugula, also known as rocket, has been grown in the United States for centuries for use as a salad green and herb, especially in Italian cooking, but it did not begin appearing in supermarkets and on restaurant menus until the 1980s, and then mostly in upscale places. In 2007 it was still unfamiliar to many Americans and those who were familiar with it generally considered it gourmet, especially in comparison to iceberg lettuce or spinach.[11] Obama may have been trying to commiserate with the rural farmer plagued by stagnant income and rising grocery bills, but his remarks were widely perceived as an impolitic reminder that he was the kind of person who could afford to shop at Whole Foods.

Washington Post columnist George Will cited the arugula quote on ABC's *This Week* when he denounced Obama for being "elitist."[12] *Wall Street Journal* columnist John Fund referred to it as evidence of Obama's "condescension towards salt-of-the-earth Democrats."[13] Tulsa-area blogger Michael Bates wrote about the incident under the headline, "Typical Liberal Arugulance," which conservative pundit Michelle Malkin repeatedly linked to on her blog in an attempt to turn the phrase into a meme.[14] Even Obama's supporters generally interpreted the arugula comment as an embarrassing gaffe. In a *Huffington Post* article titled "Obama Eats Arugula," progressive law professor Joan Williams suggested that Obama and his campaign ought to "recognize the ways Obama is sending out alienating signals of class privilege in an entirely unselfconscious way."[15]

Those who defended Obama's comments largely did so by attempting to recast arugula as a humble, accessible ingredient. One respondent wrote, "I'm not rich and I love arugula! I also know what bruschetta is, know what a panini press is, and risotto is. The problem with a lot of Americans is that they have no sense of culture,"[16] paradoxically reinforcing the association between arugula and an implicitly hierarchical notion of culture even while denying the connection between culture and wealth. Another replied: "They sell that stuff at the Olive Garden, which is not exactly upscale dining. We are not talking Grey Goose vodka and crème brulee here. We are talking about salad and bread," apparently attempting to distance arugula from truly upscale foods.[17] Several other participants in the thread noted that arugula was part of traditional Italian peasant cuisine because it grows "like a weed," again attempting to reframe it as a low-status food. One commenter speculated that you can probably even buy it at Walmart.[18] Obama also attempted to use this line of reasoning to turn the accusations of elitism on his critics. Addressing a crowd in Independence, Iowa later that year, Obama said, "All the national press, they said, 'Oh, look at Obama. He's talking about arugula in Iowa. People in Iowa don't know what arugula is.' People in Iowa know what arugula is. They may not eat it, but you know what it is."[19] He also tried to distance himself from gourmet foods. At a sports bar in Pennsylvania, he inquired about the local brew before ordering it: "Is it expensive, though? . . . Wanna make sure it's not some designer beer or something."[20]

Despite those efforts, the arugula incident followed him throughout the campaign. Almost a year after the rural issues forum, a *Newsweek* cover story titled "Obama's Bubba Gap" was represented by cluster of leaves labeled "*(a-ru-gu-la)*" counterbalanced by a frothing mug of golden "*(bîr)*." In an introductory note, the issue's editor referred to the popular impression that Obama was out of touch with common voters as "the arugula factor."[21] The common misconception that he had expressed some special affinity for arugula was reinforced by other anecdotes suggesting that he might have personal proclivities for gourmet food. Blogs such as *Grub Street* and their commenters generally expressed approval of the Obamas' affection for high-end restaurants such as Topolobampo and Spiaggia in Chicago, citing them as evidence of good taste.[22] The *Philadelphia Daily News* reported that during his campaign stop in their city, Obama did not eat a cheesesteak, instead sampling Spanish ham that retails for $99.99 a pound from a specialty importer at the Italian Market. An article in *USA Today* cited Obama's reported affection for caramels with smoked sea salt from Fran's in Seattle, which were included in a gift basket he received when he visited the city, as evidence of his "upscale" tastes.[23] A *New York Times* article about Obama's likes and dislikes included Fran's handmade milk chocolates in the likes list along with vegetables (especially broccoli and spinach), roasted almonds, pistachios, water, and Dentyne Ice. The *Adweek* blog *Adfreak's* assessment was that "on the whole, the faves list skews a bit more upscale than seems ideal for a candidate who's trying to strengthen his appeal to blue-collar voters."[24]

Even Obama's taste in burgers was portrayed as evidence of that he was a more sophisticated eater than his political predecessors. The introduction to a slideshow on the *U.S. News and World Report* website titled "What Makes Obama a Gourmet President" read:

> After eight years of boots and barbecue in the White House, the Obamas have introduced a gourmet atmosphere not seen in several recent presidencies. George H. W. Bush liked Chinese food from a northern Virginia neighborhood. Bill Clinton has a reputation for Big Macs. George W. Bush liked his grill. But the Obamas have instead hired their own chef and dined at the best restaurants Chicago and Washington have to offer. Even President Obama's choice of burger joints, Ray's Hell Burger in nearby Arlington, Va., only serves gourmet sandwiches.[25]

At one highly publicized visit to Ray's, he asked for "a spicy or Dijon mustard," which conservative pundits seized on. On his Fox News show, Sean Hannity recounted the incident and showed a clip from the famous Grey Poupon commercials featuring Rolls-Royce limousines, then quipped "I hope you enjoyed that fancy burger, Mr. President."[26] The other examples highlighted in the *U.S. News and World Report* slideshow included the fact that the Obamas had been spotted dining at expensive restaurants, had started a vegetable garden on the White House grounds, and had invited celebrity chefs such as Rick Bayless to cook state dinners.[27]

As vegetable gardens and burger joints are not always portrayed as gourmet, their inclusion in the slideshow helps illuminate the contemporary construction of culinary sophistication. Obama's tastes are called gourmet or upscale when they can be distinguished from the norm by price, novelty, rarity, and the promise of higher "quality." Starting a garden and eating at Ray's Hell Burger are gourmet in comparison to George W. Bush's grilling and

Clinton's McDonald's runs because the latter are more commonplace. State dinners under Presidents Bush and Clinton undoubtedly involved expensive banquet catering, but that constitutes the norm for a formal White House affair. Hiring a high-profile celebrity chef to oversee the menu, even one best known for a traditionally lower-status cuisine such as Mexican food, adds a mark of distinction. Similarly, in the debate about the meaning of arugula, the idea that the leafy green was gourmet was based on the claim that most people—especially the Iowa farmers he was addressing—would not know what it was or how much Whole Foods was charging for it. Those who attempted to argue that it was not gourmet emphasized its commonality and cheapness by referring to its prominence in Italian peasant cuisine or the fact that it can be purchased at Walmart.

The way people discuss Obama's supposed gourmet proclivities also suggests that although culinary sophistication may overlap with concerns about fatness, health, the environment, and authenticity, those are tertiary to the central feature of gourmetness: the pursuit of a superior taste experience. The problem with arugula was that it was "hoity-toity,"[28] not that it was seen as healthier or better for the environment than normal lettuce. Snubbing the Philly cheesesteak for Spanish ham and ordering a burger with Dijon mustard at an independently owned restaurant were taken as evidence of pretension because they suggest the pursuit of a distinctive taste experience, something better than the norm. This notion of betterness is a social construction, like all ideas about what kind of food is superior. Many people might prefer the taste of the traditional cheesesteak or Big Mac, but those are too common to seem gourmet.

The implication that there is some objective standard of excellence a food must meet to qualify as gourmet is reflected in a 2009 discussion thread on *Chowhound* (now simply *Chow*)[29] that began with the question "How would you define gourmet?" The original post asked people not to rely on a dictionary and to instead use their own words. The primary point of agreement among the roughly two dozen people who replied was that this was a difficult or even impossible task. Many of them declared that the word had been abused and might have had meaning in the past but had either been co-opted by commercial interests or diluted by indiscriminate popular use. Several people suggested that the word was so completely subjective and individual that it could not be defined. However, a few contributors noted that however slippery it might be, gourmet has widespread currency and pushed for a descriptive rather than prescriptive definition. One user submitted that gourmet generally refers to a specific kind of culinary excellence: "It can be good food without being 'gourmet' . . . gourmet equals the finest quality ingredients (therefore usually, but not always, expensive!) No shortcuts [. . . .] Macaroni and cheese made with velveeta or out of a box = NOT gourmet. Macaroni cheese made with hand grated pecorino cheese . . . = totally gourmet. Regular made-from-scratch macaroni cheese = not gourmet but still delicious."[30] This definition seems to reflect what most people who portrayed Obama as some-one with gourmet tastes were getting at. Culinary sophistication and the word "gourmet" refer to the pursuit of foods constructed as superior to normal, familiar, and accessible foods primarily on the basis of a specific construction of good taste signaled by things such as extra labor (e.g., "no shortcuts"), a higher price, and uncommon or exotic ingredients.

The arugula incident and conflicting responses to Obama's tastes also reveals a popular ambivalence about culinary sophistication, which might seem to be at odds with my claim that it constitutes an ideal. The associa-tion between gourmet foods and wealth runs afoul of the populist strain in American mass culture. This has particularly fraught implications in the world of politics, where certain kinds of class privilege have been coded liberal and un-American. Even self-identified foodies who aren't running for office sometimes try to distance their interest in gourmet or premium foods from food snobbery. However, most people writing in foodcentric forums expressed little ambivalence about Obama's tastes. For the writer of *Obama Foodorama* and her target audience, Obama's appreciation of trend-setting Chicago chefs such as Rick Bayless was something to celebrate.

The idealization of sophistication and America's populist strain don't merely coexist; both wax and wane, and frequently one tendency is in retreat as the other becomes more prevalent.[31] The broader ambivalence about culi-nary sophistication comes from a tension between the ascendant ideology of the food revolution and residual culinary populism. For much of the twenti-eth century, gourmet food was widely regarded with indifference, suspicion, or outright disdain in the United States. Although traces of that attitude clearly remain, the emergence of a contingent who openly admire Obama's taste proclivities attests to the influence of the food revolution. It's become increasingly normative to consider foods such as Dijon mustard and salted caramels desirable and the consumption of such things a sign of good taste.

The Ideal of Thinness

The second theme that emerges in the popular discourse about Obama and food is an interest in his body size and its relationship to his diet. The belief that thinness is superior, both aesthetically and nutritionally, is somewhat less controversial or politically vexed than the food movement's idealization of sophistication. The extent and effects of contemporary anti-fat bias in the United States are well documented. Fatness is widely associated with laziness, ignorance, lack of self-control, and excess sensuality. Fat people are less likely to be hired and promoted than thinner people with the same qualifications, they receive poorer evaluations at school and work, and they tend to earn less (even when controlling for factors such as education, years of work experience, and parents' education).[32] In surveys of health professionals, up to 50 percent of doctors and nurses admit to seeing obese patients as noncompliant, unintelligent, sloppy, dishonest, and hostile.[33] Children as young as five say they are less likely to want to be friends with fat people than thinner people and even show a bias against people who are portrayed in mere proximity to fat people.[34] So it's hardly surprising that Obama was widely praised for his fit physique and his consumption of foods that are deemed healthy because they're associated with thinness.

Paparazzi photos of him wearing only swim trunks on vacation in Hawaii elicited general admiration. *Washingtonian* magazine put the image on their cover with the headline "26 Reasons to Love Living Here: Reason #2 Our New Neighbor Is Hot."[35] When several pictures from the same vacation were published in the *Huffington Post*, comments ranged from, "I think it's great that in a nation where 60% of the population is obese, we have a new role model who is buff, intelligent and eats a healthy diet centered around salmon and rice." to "Obama is an immature dip stick! He hasn't gotten past the stage of adolescent "cool." He is simply flattering himself."[36] The possibility that his body size might reflect vanity or narcissism resurfaced in debates about his gym habits, but in general his thinness was portrayed as a sign of good health, good moral character, and good leadership potential.

His political opponents attempted to turn his thinness against him, but with little success. At a rally for the Republican candidate, Senator John McCain, California governor Arnold Schwarzenegger said, "I want to invite Senator Obama to the gym. . . . We have to do something about his skinny legs," seemingly appealing to the hard-body aesthetic of the late Cold War era when Schwarzenegger and other hyper-muscular men starred in Hollywood blockbusters such as *The Terminator* and *Rambo*.[37] Besides seeming dated, the quip failed to resonate with voters familiar with Obama's reputation for making regular exercise a priority (which also made its way into national news stories when the McCain campaign criticized him for putting the gym before his job).[38] In response to a story about Schwarzenegger's comments on the Reuters blog *Front Row Washington*, commenters universally rejected the idea that Obama should gain weight. One reader replied: "Obama can whip Arnold's thick ass any time on the basketball court" and another said, "Arnold is just a fake. Obama has a healthier look, very trim and climbs stairs with grace and strength."[39] Schwarzenegger's attempts to extend his criticism of Obama's body size to his politics similarly fell flat. In a characteristic response to Schwarzenegger's suggestion that Obama also needed to "put some meat on his ideas," one reader replied, "California is worse off financially and other ways than before Arnold came on as Gov. Where's the Beef, Arnold?"[40]

However, others expressed concerns that Obama's thinness might distance him from some voters in the era of the American "obesity epidemic." In an August 2008 *Wall Street Journal* article, Amy Chozick reported on the results of a query she had posed on the Yahoo! politics message board asking if Obama's body size would affect anyone's vote. One self-identified Clinton supporter said that yes, Obama "needs to put some meat on his bones"; another said, "I won't vote for any beanpole guy." Chozick noted that the footage of Bill Clinton rolling into a McDonald's for a Big Mac and fries, still drenched in sweat from a jog, was widely believed to have helped him connect to voters in conservative-leaning states such as Georgia and Tennessee, which have higher percentages of overweight and obese people than many blue states.[41] Chozick's article was picked up by the *Huffington Post*, where most people in the comments replied that they'd rather have a "skinny guy that's in shape" than a septuagenarian melanoma cancer survivor, referring to McCain.[42]

Obama himself defended his body size, saying "Listen, I'm skinny, but I'm tough," but he sometimes tried to distance himself from the kinds of food choices associated with thinness. At a campaign stop in Lebanon, Missouri, where he visited with voters at a restaurant called Bell's Diner, one of the first things he said upon his arrival was, "Well, I've had lunch today but I'm thinking maybe there is some pie." He ultimately ordered fried chicken, telling the waitress, "The healthy people, we'll give them the breasts. I'll eat the wings."[43] While he might have been responding in part to the pressure on politicians to show their appreciation for the local cuisine wherever they go, wings aren't especially identified with Missouri and he explicitly framed his choice as unhealthy. Choosing the fattening wings instead of the lean chicken breast may have been an attempt to counter the idea that his thin physique was evidence of an abnormal degree of self-restraint.

Although his thinness was almost unequivocally seen as a positive, the eating habits that supposedly produced it were occasionally criticized. In an inversion of the typical notion about what it means to cheat on a diet, a video of Obama handing off a brownie to an aide at a campaign stop was given the headline "How Obama Cheats on Eats at Meet 'n' Greets." The *Huffington Post* article about this sleight of hand explicitly connected it to his body size, describing the aide and bodyguard as Obama's "secret weapon in the war on love handles."[44] At other times, his healthy habits were portrayed as a worrying unwillingness to deviate from prevailing health guidelines. *New York Times* columnist Maureen Dowd criticized him for refusing a slice of chocolate cake at a Pennsylvania chocolate shop and for reportedly not liking ice cream, calling his eating habits "finicky" and "abstemious."[45] In a segment on MSNBC's *Hardball* called "Is Obama Too Cool?" *Bloomberg News* columnist Margaret Carlson said, "Sometimes you just want to tell the guy, 'Eat the doughnut.'"[46]

The coverage of Obama's thinness reveals that while thinness is unquestionably superior to fatness, attitudes toward the dietary choices associated with thinness are conflicted. The ideal eater is fit but not fastidious; he demonstrates self-control without being "too cool" to eat the occasional doughnut. The same choices that are generally seen as healthy and virtuous are always in danger of being seen as puritanical or elitist. The idea that having an aide to help him discretely dodge baked goods was a "secret weapon" in the war on fat or the fear that Obama's thinness might make him difficult for many voters to relate to all point to the idea that having the idealized body size set Obama apart from the masses. In this way, the discourse about Obama's body mirrors the concern about his tastes in food.

However, the idealization of thinness is the basis for an entirely different taxonomy of what counts as superior or inferior, elitist or populist. The handmade salted caramels and burgers with Dijon dressing that were portrayed as gourmet would both fall squarely in the fattening camp, along with more pedestrian foods such as chicken wings and brownies. Meanwhile, other items on Obama's list of likes such as MET-Rx protein bars and water might be seen as superior in terms of maintaining thinness but are too mass produced or mundane to signal culinary sophistication. There are also tensions between the ideal of thinness and the ideals of purity and cosmopolitan. In the case of purity, some of that tension hinges on different constructions of healthiness. When Obama referred to chicken breast as a healthy choice in contrast to the wings he ordered, he invoked a particular nutritional ideology that primarily demonizes foods high in fat and calories because of their presumed effect on body size. The supposed nutritional superiority of natural foods is often based on factors such as the presence of micronutrients or absence of pesticide residues and artificial flavors, colors, or sweeteners because of their presumed effect on many factors other than weight.

The Ideal of Purity

Although it's not entirely clear how the creator of *Obama Foodorama* got the impression that Obama was the "first president to have eaten organic food," she certainly wasn't alone in assuming that he was either on the natural foods bandwagon already or could be easily convinced to hop aboard. After his election, *Gourmet* magazine published a letter to the Obamas from celebrity chef and activist Alice Waters in which she offered to "help with your selection of a White House chef. A person with integrity and devotion to the ideals of environmentalism, health, and conservation."[47] The implication was that whoever had served the Bush White House could not possibly have been that kind of chef. Former White House chef Walter Scheib came forward to set the record straight, telling both *Obama Foodorama* and the *New York Times* that Cris Comerford, whom he had promoted to executive chef when he left the White House in 2005, was not only a talented cook and kitchen manager but was also committed to providing locally sourced, healthy food for the First Family and their guests. Scheib also noted that former first lady Laura Bush had insisted on organic produce. NPR food critic Todd Kliman applauded Scheib for having "the guts to stand up to Waters' inflexible brand of gastronomical correctness," despite the fact that all Scheib really did was point out that the White House kitchen was already cooking by Waters's rules.[48] The Obamas ultimately retained Comerford, but they also hired Sam Kass, who had worked for them as a private chef in Chicago and had previously established himself as a vocal critic of conventional, industrial agriculture and a supporter of local and organic food when he was executive chef at the Jane Addams Hull-House Museum.[49]

Michael Pollan's "Farmer in Chief" op-ed also weighed in with food policy advice for the president elect. Pollan called on the winning candidate to promote what he called sun-based agriculture by supporting agricultural education at land-grant colleges, creating tax and zoning incentives for developers to incorporate farmland into subdivision plans, providing grants to cities to build year-round indoor farmers' markets, requiring institutions such as hospitals and universities receiving federal funds to purchase locally grown food, and setting an example by growing and eating organic fruits and vegetables at the White House. Although the letter was published a month before the general election and was addressed to whoever might prevail in November, food and environment bloggers frequently referred to it as Pollan's "letter to Obama." Obama helped reinforce the assumption that he would be sympathetic to Pollan's argument when he referred to the letter in an NPR interview, even though he carefully refrained from committing to take any of Pollan's suggestions. Still, commenters on sites such as *Treehugger* celebrated, posting comments such as, "It makes me pretty excited that Obama has read Pollan," and "Michael Pollan should be appointed as an Obama Agricultural consultant asap. GOBAMA!!!"[50]

Since taking office, President Obama has mostly kept silent about "sun-based" agriculture, and he disappointed many supporters of organic and local agriculture by appointing people with connections to large industrial agricultural interests to key positions at the FDA and USDA. However, the Obamas did take at least one of Pollan's suggestions when they established an organic vegetable garden on the White House lawn in March 2009. The garden is largely seen as Michelle Obama's project and is associated with her "Let's Move" campaign against childhood obesity. However, she has also insisted on using only natural fertilizers and pesticides despite appeals from groups such as the Mid America Croplife Association and the American Council on Science and Health that urged her to consider using conventional crop protection products and to "recognize the role conventional agriculture plays in the U.S. in feeding the ever-increasing population, contributing to the U.S. economy, and providing a safe and economical food supply."[51]

These examples help illustrate what's at stake in the third relatively independent ideal of the food revolution, the veneration of the natural. Food revolution leaders such as Alice Waters and Michael Pollan sometimes argue that eating fresh, organic produce is more pleasurable than eating processed food (though certainly many Americans would not agree). However, in their public appeals to Obama, they focus primarily on social and environmental benefits. Waters's letter claimed that "supporting seasonal, ripe delicious American food would not only nourish your family, it would support our farmers, inspire your guests, and energize the nation."[52] Pollan's letter argued that sun-based agriculture will combat climate change, help the United States achieve energy independence, and improve regional economies. As a marketing term for which there is no official definition, "natural foods" usually encompasses anything organic-certified and food produced without the use of hormones, antibiotics, and artificial sweeteners, colors, or flavors. When Waters claims that "seasonal, ripe delicious American food" will "nourish" the Obamas, she invokes a sense of health that goes far beyond body size. Natural foods are sometimes supposed to prevent a whole range of ills, from hyperactivity and intestinal distress to depression, fatigue, and cancer. As Pollan argues, they are also generally seen as being more environmentally sustainable, usually because they are presumed to use fewer or less toxic agricultural inputs. Many are also seen as being more humane to animals and as promoting a better standard of living for farmers.

I use the term "purity" to encapsulate this ideal both because of its historic use by the pure food and drug movement to refer to food free from adulteration or contamination and because of its moral connotations. Instead of being primarily about taste or thinness, natural foods evoke a kind of righteousness. This seems to be what Kliman was chafing against in his accusation that Waters represents an "inflexible brand of gastronomical correctness." Kliman's reaction also captured the popular ambivalence toward this ideal. Kohan may have celebrated the (probably false) idea that Obama was the first president to eat organic food, but others—including the McCain campaign—claimed that his affection for Honest Tea, an organic-certified version of Lipton that carries the slogan "Nature got it right. We put it in a bottle," was yet more evidence of elitism.[53] Kohan and her readers were also thrilled that on the Obamas' date night in Manhattan, they chose to eat at locavore landmark Blue Hill. However, his reference to Whole Foods—the world's largest natural foods retailer—seemed to increase the perceived snootiness of the arugula incident. Like the anxieties about snobbery in relation to the ideal of sophistication, the controversy over Obama's association with foods that represent the ideal of purity is less about whether natural foods are actually superior (which has become a form of popular common sense) than what they say about his class status.

The Ideal of Cosmopolitanism

The last of the four central ideals invoked in the discourse about Obama and food is essentially adventurousness—a willingness to seek out ethnic and regionally marked foods and demonstrate an appreciation for diverse culinary experiences. Obama earned a reputation for having diverse tastes by naming Topolobampo and Spiaggia as two of his favorite restaurants (the first claims to serve "authentic Oaxacan" and the latter "authentic Italian") and by writing in The Audacity of Hope about his love of soul food and enthusing about Hawaiian specialties such as shave ice and plate lunch during the same vacation when he was photographed shirtless.[54] Geographical range isn't the only indicator of cosmopolitanism. NPR's Kliman praised the "eclectic" taste the thousand-bottle wine cellar at Obama's Hyde Park home represented, on the one hand, and his appreciation for delivery pizza and burgers, on the other.[55] Similarly, an Esquire slideshow called "Eat Like a Man" claimed that it had been "established that President Obama has pretty good taste" and then cited as evidence the claim that "the First Family is known to take food seriously, whether it's stopping into a local New Orleans gumbo shop or celebrating Michelle's birthday at Nora Restaurant in D.C."[56] Perhaps because of all the attention his upscale tastes attracted, many of the incidents that were seized on as evidence of Obama's adventurousness involved foods with a little more populist cred.

For example, both news media and blogs praised his first public visit to a Washington restaurant, which involved meeting District of Columbia mayor

Adrian Fenty at Ben's Chili Bowl, a restaurant with deep cultural and historical significance for the city and the African American community. Ben's is
equally famous for its chili-smothered sausages called half-smokes and for
serving musicians such as Duke Ellington, Nat King Cole, and Ella Fitzgerald
when they used to perform at neighboring U Street venues. It also served as
a gathering place for community organizers during the civil rights era. Food
blog *Chomposaurus* covered Obama's visit to Ben's in an entry subtitled, "Our
New President Knows How to Eat,"[57] and on *Obama Foodorama*, Kohan
said, "We kinda *adore* the fact that Barack's first public restaurant outing in
Washington was to a joint that's entirely *of the people*, and also very historic."[58]

As much of the coverage also noted, Obama's visit was prefaced by a
change in the landscape of the restaurant. For years, a sign hanging behind
the counter at Ben's read: "List of who Eats Free at Ben's / Bill Cosby / NO
ONE ELSE /—MANAGEMENT," but the day after Obama was elected, the
management hung a new sign reading: "Who Eats Free at Ben's: *Bill Cosby /
*The Obama Family." The MSNBC program *Meet the Press* invited Cosby to
participate in a round table on the day of Obama's visit to Ben's, and upon
hearing that Obama had asked for an explanation of what a half-smoke is
when he got to the counter, Cosby joked, "I'm taking my vote back." He also
suggested that while the First Lady and her mother might deserve free eats
at Ben's, the president himself hadn't earned the privilege.[59] Cosby's ribbing
played on anxieties about Obama's racial identity: Was he really black? Was
he black enough? However, it was actually Obama's failure to pass completely
that made his trip to Ben's newsworthy. Knowing how to order a chili dog was
not seen as an intrinsic part of Obama's identity. If it had been, his trip there
would not have been notable. Eating at Ben's crystallizes the quintessential
marker of the ideal of cosmopolitanism: a willingness to seek out novel or
exotic culinary experiences, foods specifically marked as Other to one's own
identity. Adventurous eating is particularly signaled by the willingness and
ability to appreciate both sophisticated foods and humbler ones distinguished
by their imagined connection to history, geography, or tradition.

Not everyone was impressed by Obama's visit to Ben's, and an article by
Alicia Villarosa in the online magazine *The Root* exemplifies the tensions
between the four ideals of the food revolution. Villarosa specifically juxta-
posed his trip to Ben's with the "healthful" choices she says he and Michelle
should adopt both for their own sake and to "set the tone for our nation." She
writes, "We know the new president likes half-smokes from Ben's Chili Bowl
and his mother-in-law's sweet potato pie, but how much does he eat veggie-
centered meals? Stocking the White House kitchen with organic food could
help, too. Even though in these tough economic times many Americans cannot

shop organic, nudging people toward food full of natural nutrients minus the chemical pollutants would be an enormous boost to our national health."[60] Instead of celebrating his trip to Ben's as evidence that Obama "knows how to eat," Villarosa suggests that authentic, racially marked foods ought to take a back seat to a conception of health rooted in the ideal of purity. Similar comments followed the media coverage of his visits to DC-area burger joints. In June 2009, the DC insider newspaper *Politico* featured as their question of the week: "Are Obama's burger runs a bad example to be setting for America's citizens?" The online version included a video of musician and fashion mogul Russell Simmons suggesting that Obama look into vegetarian options such as falafel, which he suggested would be healthier.[61]

In a seeming reversal of her previous critique of Obama's "abstemious" refusal to eat cake, Maureen Dowd suggested that Obama should stop using burger runs to "beef up his average Joe image." Then, referring to a special on health care that Charlie Gibson and Diane Sawyer were scheduled to broadcast from the White House the following week, she recommended that Obama "forgo the photo-op of the grease-stained bovine bag and take the TV stars out for what he really wants and America really needs: some steamed fish with a side of snap peas."[62] Whether his burger runs were "fancy," as right-wing pundits claimed, or a kind of populist posturing, these critiques point to the gulf between ideals based on taste and authenticity and ideals based on thinness or purity. As an article about the growing popularity of Obama's favorite candy also noted, "One thing salted caramels conspicuously lack is a health and wellness angle."[63] Just imagine how Cosby might have reacted to Obama's ordering anything comparable to steamed fish and snap peas at Ben's.

Sophistication and cosmopolitanism aren't always aligned, either. Topolobampo and Spiaggia might exemplify both, but the elevation of Dijon mustard has less to do with its foreignness than its claim to superior taste. The foreign, Italian peasant associations with arugula were invoked in an attempt to make it seem less gourmet, and however delicious or authentic a half-smoke or Hawaiian plate lunch might seem, they don't involve the expensive or rare ingredients or labor-intensive preparation usually required to signal culinary sophistication. The one thing that unifies the four relatively independent ideals is the ambivalence about them their association with the professional middle class provokes.

NOT ENLIGHTENED, BUT ELITE

The conflicts between the four ideals challenge the notion that there's any true consensus about what it means to eat well. Thus, they also undermine the prevailing explanation for the proliferation of popular interest in and anxiety about food in the last few decades. More knowledge and access didn't lead people to choose inherently tastier, healthier, more ethical, and authentic foods because there's no real agreement about what foods fulfill any of those ideals, let alone all of them at once. Obama was both praised and criticized for liking everything from upscale Mexican food to chili-covered hot dogs, from organic backyard vegetables to burgers with Dijon mustard. Given the range of foods that apparently signify good taste, it might appear as though what increasing knowledge and access has created is instead a great diversification of ways to eat better. However, the four ideals don't represent some random proliferation of culinary preferences. Instead, as the popular discourse about Obama and food also reveals, the unifying characteristic of the food revolution's otherwise incompatible ideals is their association with the elite.

The arugula incident and concern that Obama's upscale tastes might distance him from the average voter attest to the link between foods portrayed as sophisticated and the elite, which makes sense given that a higher price is one of the features that distinguish gourmet foods. Similarly, the *Wall Street Journal* article suggesting that that Obama's thinness might distance him from the masses reflects the current association between thinness and wealth.[64] A *Gawker* post about the shirtless vacation photos made the link explicit: "Barack Obama Shames Americans with His Elitist Body."[65] Organic foods are typically more expensive than their conventional counterparts. Last, the appreciation for diverse cuisines is often seen as the prerogative of elites because it typically requires mobility and capital. Obama's visit to Ben's Chili Bowl might have been notable primarily for its association with the masses, but his failure to pass completely made his visit there a kind of slumming, the quintessentially elite practice of dabbling in lower-status consumption.[66] Slumming ultimately reinforces class hierarchies by affirming that the elite patron doesn't really belong in the slum and has the ability to leave.

However, wealth isn't the only factor that distinguishes the foodie elite. After all, Hillary Clinton and John McCain are fantastically wealthy senators, but their food choices weren't subject to anywhere near as much hopeful projection, scrutiny, and debate in 2008. Instead, most people seemed to assume that their tastes were unremarkable. The same is true of most of the 2016 presidential hopefuls. The extra attention lavished on Obama may have been due in part to his relative youth. McCain graduated from college in the late 1950s and Clinton in the 1960s, before the four ideals became mainstream again. Obama graduated from college in 1983, just as trends such as gourmet cooking, dieting to lose weight, buying natural food, and eating at ethnic restaurants were really taking off. It might also have had something to do with the fact that he made Chicago his hometown. The food revolution has always been associated more with the coasts and urban centers than with places such as Arkansas and Arizona. But probably the most important factor was the general perception that he was further to the left ideologically than either his Republican or Democratic competitors. Just as George W. Bush's political conservatism probably influenced Alice Waters's assumption that the White House chef in his employ would not prioritize organic and local produce, the fact that Obama was seen as a liberal probably helped convince many people that he must be an enlightened eater.

Natural food has been associated with leftist or radical politics since the late 1960s. As Warren Belasco chronicles in *Appetite for Change: How the Counterculture Took on the Food Industry*, ecology emerged as a "fresh oppositional alternative" for burned-out and disaffected hippies and activists in the aftermath of the King and Kennedy assassinations and the 1968 Chicago Democratic convention. Many of them turned to vegetarianism, whole grains, and gardening as a new, more personal and tangible form of protest and self-enhancement.[67] The further association with the elite didn't occur until the 1980s, when the food industry found that it could appeal to the upscale sector of the fragmenting middle market by invoking aspects of what Belasco calls the "countercuisine," after its association with the 1960s radicals. As granola, sprouts, yogurt, and tofu made their way into ordinary supermarkets, they were particularly associated with a new demographic: the yuppie.

Yuppies, or young urban professionals, were the target of widespread scorn and criticism from essentially the first moment they were given a name, even though the population that actually fit the description was tiny. One marketing research group estimated that the population of 25- to 39-year-olds earning more than $40,000 in professional or managerial positions in 1985 was about 4 million, of whom only 1.2 million lived in cities.[68] Nonetheless, Belasco argues that market analysts and food manufacturers took them seriously because they saw them as taste leaders: "According to Market Facts, a Chicago-based research company, the 'true yuppies' were just the tip of an iceberg composed of people who watched 'Cheers,' 'St. Elsewhere,' and 'Hill Street Blues' and aspired to grind fresh coffee beans, drink imported beer or wine, own a personal computer, and use automatic teller machines."[69] As this description also suggests, yuppies were associated with far more than just natural food. Actually, all four ideals of the food revolution were used to define the habits of these emerging taste leaders. The woman on the cover of *The Yuppie Handbook*, a tongue-in-cheek field guide published in 1984, carries a gourmet shopping bag (see Fig. 2). Both pictured individuals are slim, and their tennis shoes and the man's squash racquet nod to their participation

Figure 2. *The Yuppie Handbook.* Source: Marissa Piesman and Marilee Hartley,
The Yuppie Handbook (New York: Long Shadow Books, 1984). Cover design
by Jacques Chazaud.

in the growing fitness trend that accompanied the increasing idealization of
thinness. The fresh pasta in the gourmet bag and references to tortellini and
tuna sashimi in the manual evoke the ideal of cosmopolitanism.[70]

New York Times columnist David Brooks coined another term for the
group of people distinguished by exotic, upscale consumption habits and
liberal attitudes toward sex and morality in his 2000 book *Bobos in Paradise*:
"Bobo," a portmanteau of bourgeois and bohemian.[71] According to Brooks, the

central tension in Bobo culture is how to reconcile egalitarian ideals with class pretensions. The primary strategies Bobos use are 1) investing deeply in the myth of meritocracy, which enables them to believe that their status is the result of skill and effort rather than unearned privileges; and 2) using their consumption choices as opportunities to express their political beliefs by fetishizing products that are labeled "Made in America," Fair Trade certified, or organic or are otherwise perceived as socially responsible to express their distinction from the established mainstream. The problem, Brooks says, is that their attempts to distinguish themselves from the mainstream ultimately seem hypocritical. They claim to believe in equality yet spurn the tastes of the masses. Politically conservative elites might also engage in conspicuous consumption, but as they are less invested in egalitarianism, such behavior is less likely to raise populist hackles. As journalist Mark Ames said in an essay about the 2004 presidential election, "Republican elites don't set off the spite gland in the same way."[72]

The problem of the liberal elite is the inverse of the problem of the culturally conservative working class that Thomas Frank describes in *What's the Matter with Kansas?* According to Frank, over the second half of the twentieth century, the Republican Party attracted white working-class voters by appealing to cultural wedge issues such as abortion and guns, thus inducing them to vote "against their interests."[73] For the liberal elite, the question could be rephrased: "What's the matter with Manhattan?" Why would many people with disproportionate wealth and power support policies that appear to be against their economic interests? And why do they do it while simultaneously flaunting their privilege by buying things such as gourmet food? It's telling that the criticism of the liberal elite across the Anglophone world often explicitly invokes high-status foods. In Australia, the term "chardonnay liberal" emerged in the late 1980s to describe the "guilt-ridden rich and bleeding hearts" who supported the left-leaning Labor Party.[74] In Britain "champagne socialist" has the same connotation, and in 2002, Ireland's Labour Party leader Ruairi Quinn was derided as a "smoked salmon socialist."[75]

The real mystery in both cases isn't why people vote against their economic interests—clearly, people often prioritize other issues—but why the same tastes that make the liberal elite the subject of ridicule have nonetheless become increasingly normative. The market researchers' suspicion in the 1980s that yuppies were taste leaders has been borne out by the growing popular embrace of the four ideals of the food revolution. Many trends that *The Yuppie Handbook* portrayed as bizarre and comical in 1984 have become far more widespread. Referring to the handbook in a 2006 article for *Details* magazine, Jeff Gordinier says, "The yuppie could be found working off stress with a shiatsu massage and a facial, learning as much as possible about fine wine, traveling around the world on vacation . . . and—the clincher—eating tuna sashimi for lunch! . . . All of which means that the archetypal yuppie of the eighties sounds precisely like, um, *everyone you know.*" If anything, he says, the average middle-class consumer today would make the trendiest yuppies feel provincial: "No host worth his fleur de sel would serve brie at a cocktail party—not when there are hundreds of obscure cheeses on display at Trader Joe's."[76] As Gordinier suggests, some of the details may have changed, but the ethos of yuppie consumption—the pursuit of upscale, diet, natural, and exotic foods—has not only persisted but spread.

In part, the mainstream success of what began as yuppie tastes probably has something to do with that demographic's disproportionate influence on the cultural texts that shape popular beliefs. The culture industries—major movie and music studios, book and magazine publishers, television broadcasters, and advertising agencies—are concentrated in urban, coastal areas. Virtually by definition, most journalists, university professors, poets and novelists, filmmakers, television writers, comics, musicians, and people involved in advertising, design, education, and the arts are part of the professional middle class. The creative industries are also generally seen as being more liberal than other professional fields such as engineering, medicine, law, finance, and management.[77] As the urban liberal elite embraced the four ideals of the food revolution, those ideals were reflected in everything from news reports on emerging food trends to representations of food and eating in movies to the design of food packaging.

Additionally, as much as yuppies have always been subject to ridicule, they still represent an enviable position in the American class hierarchy. In Paul Fussell's irreverent 1983 analysis of social hierarchies in the United States, *Class: A Guide through the American Status System*, he argues that there are nine distinct classes and that everyone, even the truly rich, really want to be upper-middle class.[78] Fussell doesn't explain why that is, but I suspect the key is that the upper-middle class isn't just wealthy but also because they're seen as upwardly mobile. As a subset of the upper middle class that might not actually have as much capital or power as more established professionals or people with inherited wealth, yuppies weren't so much positioned atop the social ladder as they were poised for success. They were a symbol of being up and coming, which positioned them perfectly to become the cultural repository for fantasies about class mobility. Furthermore, since their status was defined more by their consumption habits than their actual wealth or income, being like them could be achieved simply by consuming like them.

How Taste Discriminates

The idea that consumption practices communicate important information about status and thus might be manipulated for personal gain isn't new. In 1899, Thorstein Veblen introduced the term "conspicuous consumption" in *The Theory of the Leisure Class* to describe the behavior of the nouveau riche who had profited from America's nineteenth-century industrial revolution. According to Veblen, their acquisition of goods such as silver utensils and luxury automobiles were driven by the desire to display social power and gain prestige, not by any real differences in the quality or utility of the expensive products they favored. Veblen was critical of conspicuous consumption, arguing that the pursuit of social capital and ego protection were not useful purposes and were ultimately illogical.[79] We all need friends and allies to survive, and social esteem is crucial to our life chances, but Veblen is hardly alone in looking down on explicit bids for status. No matter how explicable or deeply rooted in human social behavior it might be, status climbing is still widely seen as venal or crass.

It's not only the rich or people with surplus income whose consumption habits and preferences are governed by status concerns. Perhaps the most extensive and influential analysis of how status shapes tastes, behaviors, and consumption practices for people of all classes is sociologist Pierre Bourdieu's 1979 book *Distinction: A Social Critique of the Judgment of Taste.* Bourdieu argued that everything from what kind of music people enjoy to what kind of photographs they like to hang on their walls, what kinds of hobbies they pursue in their leisure time, and what kinds of food they like to eat are governed by two primary factors, both of which are related to class: their economic and educational background and their quest for status. When Bourdieu and his team of researchers gave 1,217 French people drawn from all different social classes an extensive questionnaire about their preferences in the 1960s, they found distinctive "systems of dispositions (*habitus*) characteristic of the different classes and class fractions."[80] In short, poorer people tended to prefer lowbrow things, middle-class people liked middlebrow things, and rich people liked highbrow things. For example, when asked whether a series of subjects would make interesting or beautiful photographs, manual workers overwhelmingly preferred a sunset or social event such as a first communion to random objects such as the bark of a tree, a snake, or pebbles, which they described as a waste of film or dismissed as bourgeois photography. The wealthier and better educated respondents were far less excited about the sunset and social events and saw far more artistic potential in the random objects.[81]

Many of the preferences the survey recorded reflected fine divisions within classes: craftsmen had different tastes than small shopkeepers, who in turn were different from commercial employees although they were all similar in terms of income, and the same was true of engineers compared to executives. However, the general pattern reflected in every area of culture Bourdieu's team asked about was that the lower classes reported preferences that were "identified as vulgar because they are both easy and common" while the higher classes favored "practices designated by their rarity as distinguished."[82] According to Bourdieu, the latter are not seen as elevated because they are objectively better or universally pleasing. On the contrary, "natural" enjoyment and all the things that most people find viscerally pleasing were typically seen as "lower, coarse, vulgar, venal, servile."[83] The reason the upper-class preferences work to set people apart, or distinguish them, is precisely because they depend on scarcity, or knowing and liking things that most people usually don't.

The social nature of taste doesn't mean that people always consciously choose what to like based on what they hope it will make people think about them. Although Bourdieu's respondents demonstrated that people are highly aware of the social significance of their tastes, their habitus was also shaped by their upbringing, their education, and the social structures they inhabit. Some working-class people might reject bourgeois photography as a conscious expression of their own class identity (i.e., they dislike abstract art specifically because they see it as elitist or they embrace folk art in solidarity). However, many are skeptical about or indifferent to art for art's sake because it doesn't reflect the kind of art they've learned to like and their beliefs about what is beautiful. Similarly, the richer people aren't necessarily just saying that they see the beauty in the photographs of random objects because that's what they know is supposed to be classy; it may be what their upbringing and education has truly predisposed them to find interesting and attractive. People's tastes are based partially on their deliberate attempts to perform a particular status or gain some competitive advantage and partially on the spontaneous, visceral attractions and revulsions they develop based on their upbringing and education.

One of Bourdieu's main contributions was showing how the competition for the kinds of status that taste can confer, which he referred to as cultural capital, is rigged. The dominant social classes have disproportionate influence over the process of cultural legitimation, meaning they have more of a say over what counts as good taste. They tend to occupy the professions and control the institutions such as schools, museums, media industries that shape how people see the world. They tend to win more of the battles over the narratives people create to give their preferences social meaning. Thus, the dispositions of the elite get consecrated as superior instead of being seen as the predilections of wealth. Furthermore, their tastes are often difficult for people without an elite socioeconomic background to acquire.

In the decades since *Distinction* was published, scholars have debated the extent to which Bourdieu's theory still applies or whether it ever applied anywhere outside 1960s France. Some sociologists argue that taste hierarchies have broken down and that we've entered an era of omnivorousness in which standards have democratized. When it comes to food in particular, the elite have embraced multicultural cuisines, and gourmet foods are far more accessible to the middle and working classes than they were in the eighteenth and nineteenth centuries.[84] However, these supposedly omnivorous developments may function more as a new way of differentiating between classes than as a real relaxation of taste hierarchies. Richard Peterson argues that in place of traditional highbrow/lowbrow divides, high status is now marked by the ability to draw on multiple cultural forms. So, for example, instead of exclusively listening to classical music and opera, the elite today acquire and perform cultural capital by being knowledgeable about a variety of musical genres, from hip-hop to folk to jazz.[85] As richer people tend to have more access to a variety of cultural realms, it's much easier for them to cultivate a diversity of tastes.

Whether these developments signal a real move toward democratization or merely a new kind of class distinction, it's important not to overstate the extent of the new omnivorousness. As in music, where many popular genres such as heavy metal, country, and diva pop are still largely excluded from critical acclaim and fail to signal elite cultural capital, even if you sample widely from all types of food, many of them are excluded from the contemporary construction of good taste. Eating, enjoying, and displaying extensive knowledge about the food served at national chains such as McDonald's and Applebee's, products such as Hot Pockets and Lunchables, or the kind of food served at institutions such as low-income public schools, prisons, and nursing homes will not earn you elite culinary capital (a term coined by Peter Naccarato and Kathleen LeBesco to refer to the ways food and food practices mark social status), no matter how diverse your preferences within those realms are. In practice, high-status omnivores are still pretty picky eaters.

When people turn their nose up at low-status foods such as McDonald's and Hot Pockets, it isn't necessarily because of their low status, at least not directly. Similarly, when they embrace high-status foods such as arugula or artisanal salted caramels, it isn't necessarily because those things carry elite culinary capital. The relationship between taste and status is frequently

mediated by ideology. In calling attention to the role of ideology, I'm depart-
ing slightly from Naccarato and LeBesco's explanation of how culinary capital
works. They describe people's use of the symbolic power of food as a process
of deliberate, strategic self-creation: "Individuals begin to govern themselves
by choosing to adopt practices and behaviors because of the status that comes
with doing so."[86] That description may capture the basic cultural logic of culinary capital: food and food practices serve as markers of social status, so food
serves as a kind of symbolic vocabulary that people can use in an attempt to
perform the status they want to be. However, it makes the cultivation and
deployment of culinary capital sound a lot more deliberate and intentional
than it usually is.

Sometimes people consciously and deliberately select foods that they
believe will communicate a particular kind of status for that reason, but that
kind of deliberate status-seeking performance is actually pretty perilous. If
you're suspected of selecting food primarily because of its elite connotations,
you risk being called a snob, which is a way of saying you're a status fraud. The
word "snob" pulls the rug right out from under any attempt to climb the class
ladder. Most of the time, people don't choose high-status foods because of
their association with the elite. Instead, they believe those foods are actually
better—better tasting, healthier, better for the environment, more authentic,
and so forth. The accumulation of culinary capital is indirect. The reason
tastes are a reflection of class is because people's class status shapes their
investment in the beliefs that guide their consumption choices.

For example, consider the rise of what historian Helen Veit calls the "moral
imperative of self-control around food" that arose during the First World
War.[87] In 1917, shortly after the United States entered the war, the government
created an agency called the U.S. Food Administration with the goal of providing supplies for the troops and civilians in Western Europe, where provi-
sions were sometimes desperately scarce. When the Food Administration
called on Americans to voluntarily consume less of high-calorie, easily
transportable staples such as beef, pork, white flour, butter, and sugar, many
Americans "welcomed the opportunity . . . to exercise self-discipline."[88] Both
the bureaucrats in the new agency and many ordinary Americans began
to denounce overeating, waste, and the sensual pleasures of eating as they
elevated asceticism and rational decision making about food based on the
emerging science of nutrition. People actually clamored for stricter rules.
Even though the Food Administration never exercised the power Congress
gave it to make rationing compulsory or to requisition desirable commodi-
ties for export, many citizens requested that they do so. In 1918, a group of
wealthy New York women wrote to the Food Administration demanding

that the government tell them what to eat, and the Food Administration responded by printing a run of special ration cards setting strict weekly limits on red meat, butter, wheat flour, and sugar. The enthusiasm for rationing wasn't limited to the elite, either. According to the Food Administration's records, about 70 percent of American families signed pledge cards saying they would comply with the recommended food conservation measures.[89]

However, not everyone was entirely eager to sign the pledge cards, and some of the "bitterest objections" had to do with their "incompatibility with the economic realities facing poor Americans." As Veit notes, "the choice to live more ascetically was a luxury, and the notion of righteous food conservation struck those with nothing to save as a cruel joke."[90] A letter to Herbert Hoover from a southern farmer's daughter noted that their meals were always wheatless, meatless, and butterless by necessity, and an editorial in a Georgia paper quipped: "Lots of us would like to be 'rationed' very frequently with chicken and rice, corn bread, hominy and milk 'as much as we could desire,'" referring to the substitutions for red meat and wheat bread the Food Administration recommended.[91] This example clearly illustrates the material restrictions of class on the use of culinary capital: poor people couldn't participate in a meaningful way in the rationing program, which was moralized and seen as a way of performing good citizenship during the First World War, because they couldn't even afford the foods that were being recommended as substitutes, let alone the ones that richer people gave up, thus demonstrating their superior self-control. It also illustrates how social class can shape the embrace or rejection of particular ideologies.

The dominant meaning of voluntary self-restriction and asceticism in the early twentieth century depended on access to abundant and varied food. For people with access to plenty, dietary restriction was available as a sign of self-mastery and maturity and even freedom. They could see choosing frugality as evidence of their ability to forgo immediate self-gratification and rein in the baser desires of the body. On the other hand, for immigrants barely subsisting on bread, coffee, and watery noodle soups, further restrictions and rationing would appear to be the very opposite of freedom. As Bourdieu described in *Distinction*, for the poor and working class, the ideal meal is characterized by plenty above all else. Particularly on special occasions, the impression of abundance is paramount and meals consist of what Bourdieu described as "elastic" dishes: soups, sauces, casseroles, pasta, and potatoes served in heaping dishes brought directly to the table with a ladle or spoon to avoid too much measuring and counting. Everyone is encouraged to serve themselves and men are often expected to take second helpings. For the bourgeoisie, in contrast, a desirable meal would consist of plated courses or at the very least individual portions of fish, meat, vegetables, and dessert served with much more ceremony, and there would be no seconds. According to Bourdieu, both of these seemingly disparate ways of eating represented a kind of freedom. For the working class, plenty is the freedom to have enough and freedom from stuffy or pretentious ritual; for the bourgeoisie, formality is freedom from "the basely material vulgarity of those who indulge in the immediate satisfactions of food and drink."[92]

These antithetical beliefs and practices get associated with the different classes who tend to ascribe to them, and the ones associated with the middle class or elite carry higher-status cultural capital. The Americans who participated in rationing were not necessarily engaging in a deliberate quest for status. The fact that voluntary rationing was a way for them to distinguish themselves from the poor and cement their class status probably influenced their embrace of self-control and asceticism and helped make that ideology mainstream. But it's the belief that rationing was morally superior itself that would have enabled them to make sense of their behavior, not the status concerns that may have inspired them to embrace that ideology. Although replacing wheat with corn and eating meatless meals served as a form of middle-class cultural capital during the war, most people who engaged in it probably did so because they thought it was the right thing to do.

A more recent example that demonstrates how this works on the level of taste is the rise of kale as both a trendy ingredient, especially on blogs and on menus in restaurants that follow a farm-to-table aesthetic, and as a kind of avatar of the food revolution and the urban elite lifestyle that movement is associated with. The popular obsession with the hearty cooking green has prompted a number of people to ask Why kale? Most attempts to answer that question point to kale's nutritional content, but kale isn't a significantly better source of most micronutrients than other green vegetables such as spinach or broccoli or Brussels sprouts. A report by the Centers for Disease Control ranked kale number 15 on a list of the 47 most nutrient-dense fruits and vegetables, assigning it 49 points out of 100 based on 17 "nutrients of public health importance," well below the score of many other leafy greens, including collards (62 points), romaine (63 points), spinach (86 points), Chinese cabbage (92 points), and watercress (100 points).[93] Instead, what probably distinguishes kale is that it tends to be the bitterest of the bunch and thus hasn't been as popular as other leafy greens and brassicas in American cooking. Its relative obscurity, at least before the mid-2000s, made liking kale more of a mark of distinction than liking spinach or broccoli. Kale had a sort of symbolic availability, but I suspect that its bitterness is also part of its appeal.

Bitterness is a taste sensation that most people have an inherent aversion to as children but that many people develop a fondness for as an adult—just as they do for spicy, astringent, and pungent flavors. Kale's bitterness means that developing a taste for it usually requires repeated opportunities for exposure and some kind of incentive, and both of those things are affected by class in ways that predispose middle- and upper-class people—particularly the politically left-leaning and urban ones—to cultivate an affection for it while preventing most poor or working-class people from doing the same. The nutritional claims that have earned kale superfood status are more likely to appeal to the members of the white middle class who have embraced the ideology of health as achievement that Robert Crawford calls "healthism."[94] People who are less preoccupied with personal health or less convinced that their health depends primarily on their lifestyle habits (as opposed to genetics, luck, or structural factors such as poverty) will generally be inclined to make it a point to eat foods that are advertised as nutrient dense. Kale was also popularized by the rise of community-supported agriculture and farmers' markets, particularly in regions with short growing seasons, because it's easy to grow and frost tolerant and because those direct-to-consumer produce marketing services tend to serve a disproportionately wealthy population.[95]

Perhaps most important, the exotic or novel nature of kale (at least before the mid-2000s) would probably be an asset to middle-class consumers with an interest in distinguishing themselves from the masses and a deterrent to poor and working-class consumers who risk more and gain less from culinary adventurousness. As Reay Tannahill writes, "The nearness of hunger breeds conservatism. Only the well-fed can afford to try something new, because only they can afford to leave it on the plate if they dislike it."[96] Additionally, there are risks in eating foods associated with higher-class status, especially if you aren't a member of that group. There may be rewards for learning to eat, like, and talk about high-status foods in the right social context as long as it's interpreted as genuine, but if it comes off as pretentious and gets read as food snobbery, it just invites scorn. None of that means the people who eat kale are doing so purely because of the cultural capital it has come to represent. On the contrary, most of them have probably cultivated a genuine affinity for the way it tastes, bitterness and all, just like I cultivated a genuinely preference for black coffee and many people come to like dry red wine and India pale ales. But the beliefs that lead people to cultivate those tastes and the rewards they get for doing so are shaped by their class background and status aspirations, just as Bourdieu argued.

One thing Bourdieu's theory of cultural capital doesn't explain is how and why culturally dominant tastes and the ideologies people use to make sense of them change over time. *Distinction* didn't set out to explain shifts in taste; it was based on a snapshot of French people's preferences at a single moment in the 1960s. However, the common theme of anxiety about elitism that unites the discourse about Obama's relationship to each of the four ideals reveals how class might also play a role in cultural transformations such as the food revolution.

The fact that these relatively independent ideals became mainstream at the same time as income inequality began to increase and class mobility stalled suggests that their appeal may have had more to do with their ability to confer status than with people's knowledge about or access to better food. This possibility is supported both by the pervasive anxieties about class and status in mass media representations of food and the longer history of American beliefs about superior taste. Although the foods associated with the yuppie or liberal elite seemed new in the 1980s, that's only because they presented a dramatic contrast to the dominant tastes of the 1950s and 1960s. The four ideals of the food revolution actually turn out to be much older, and the last times they were culturally mainstream were also periods of dramatic inequality and frustrated mobility: the Gilded Age and the Progressive Era. As the next chapter explores, the proliferation of interest in distinctive foods in those periods also served as a form of status acquisition and class distinction.

CHAPTER 2

Aspirational Eating

FOOD AND STATUS ANXIETY IN THE GILDED AGE AND THE PROGRESSIVE ERA

High breeding may be imitated, and a gentle courtesy of manner may be acquired through the same process by which other accomplishment is perfected.

—*Abby Longstreet,* Social Etiquette of New York *(1883)*

The assumption that the food revolution is something radically new is due in part to the brevity of living memory. It wasn't so long ago that grocery stores began to sell gourmet cheese—irregular hunks of Parmigiano-Reggiano with the name stamped on the rind, wedges of creamy brie, and cheddar labeled by age—usually in a separate display case in an entirely different part of the store than the rectangular bricks of Colby and Monterey Jack and individually wrapped slices of American cheese. When early Gen Xers were children, the only artificially sweetened sodas on the market were Tab, Diet Rite, and Patio. Dieters were a niche market that major brands didn't have to cater to, and no one but body builders and boxers had gym memberships. Anyone who has lived through the last three decades in America has personally witnessed the spread of gourmet kitchen emporiums, weight-loss diets, farmers' markets, and ethnic restaurants from coastal cities into suburbia and the heartland.

While many of the trends and products that exemplify the food revolution were truly new additions to the American culinary scene in the 1980s—examples include Wolfgang Puck restaurants (his first, Spago, opened in 1982), Diet Coke (introduced in 1982), Whole Foods Market (founded in Austin, Texas, in 1980), or Sriracha (created in 1980)—the desire for sophisticated, slimming, natural, and cosmopolitan foods is not new. It was not new in 1963 when Julia Child first appeared on WGBH Boston or in 1990 when the USDA created its organic certification program. It's just that most people alive today

didn't witness the last great flourishing of attempts to eat better according to
the four central ideals of the food revolution. Fancy dinner parties, dieting for
weight loss, pure food activism, and international foods were wildly trendy
from approximately 1880 to 1930. World War I caused only a brief interrup-
tion in their popularity. They were adopted first by the urban upper-middle
class, which in that era meant the 10 to 15 percent of Americans who lived
in or near cities and could afford to employ a least one full-time servant.
In practice, they may never have spread much beyond that demographic.
However, just like today, the beliefs of the urban upper middle class about
what it meant to eat better shaped the national discourse about food that was
reflected in restaurant menus and cookbooks, etiquette guides, magazines
and newspaper advice columns, advertisements, and the activities of social
and civic organizations.[1]

The particular manifestations of the ideals of sophistication, thinness, purity,
and cosmopolitanism may have changed, but the basic logic and aesthetic principles that shaped what it meant to eat better in the Gilded Age and in the last three
decades are essentially the same. Gourmet foods are supposed to taste better, at
least to people with well-trained palates, often because they contain more expensive ingredients. Thinness is both more attractive and healthier than fatness, and
it is assumed that anyone can achieve it through the right combination of diet and
exercise. Natural diets are supposed to be both healthier and morally superior,
although at the turn of the last century reformers were more concerned with
indigestion and sexual urges than with cancer and the environment. Appreciating
international and ethnic foods is supposed to be a sign of education, worldliness,
and a forward-thinking appreciation of diverse cultures. The prevalence of all of
these beliefs at the turn of the last century fundamentally undermines the idea
that today's food revolution is the result of some kind of recent enlightenment or
progress. Americans didn't just suddenly realize that gourmet food and weight
loss were desirable in the 1980s; all four ideals were mainstream a full century
earlier.

The mid-century decline of these ideals also indicates that they aren't
inevitable complements of increasing wealth and knowledge. Instead of continuing to increase in popularity the way a phenomenon driven by progress
or enlightenment presumably would have, mainstream interest in the ideals
diminished after the 1920s. French restaurants that were shuttered during
Prohibition and the Great Depression didn't reopen, even after Prohibition
was repealed and the economy rebounded. Instead, supermarkets and
chain restaurants offering standardized products spread across the country.
Advertisements began to emphasize value, familiarity, and Americanness
rather than sophistication or foreignness. The meat-and-potatoes dinner

came to dominate the mid-century national culinary imagination.[2] None of the four ideals disappeared completely, but they receded to the cultural margins.

As the Gilded Age popularity of sophistication, thinness, purity, and cosmopolitanism suggests, the recent food revolution isn't revolutionary in the sense of something radically new that opposes the prevailing order—as in a break with the past or a turn away from it—but it might be revolutionary in the sense of a cyclical rotation. The contemporary movement is a rebellion against the culinary zeitgeist that took shape during and after the two world wars, but instead of staking out new ground, it is a return to the kinds of culinary practices that prevailed before that zeitgeist got established. Many accounts of the contemporary food revolution have missed the parallels with the Gilded Age because they simply don't look back far enough; they usually take World War II as their starting point. For example, the historical overview of gourmet culture in Josée Johnston and Shyon Baumann's *Foodies: Democracy and Distinction in the Gourmet Foodscape* begins with the launch of *Gourmet* magazine in 1941 (see Fig. 3).

This chapter extends that timeline to the left, exploring a slightly longer history of attempts to eat better by drawing on what historians have written about restaurants, nutrition, consumer culture, progressive activism, fashion, and beauty in America before the world wars and their marked contrast with the popular discourse about food through the Civil War and Reconstruction eras. The chronology of these accounts is remarkably consistent. Almost all agree that the widespread interest in sophistication, thinness, purity, and cosmopolitanism first emerged in the 1880s and declined in the late 1920s. However, the causes of this emergence and retreat are a matter of some dispute. Scholars have variously attributed it to industrialization and urbanization, the expansion of U.S. empire building and the influx of immigrants, gender politics and the cult of domesticity, the growing deference to new sources of moral and scientific expertise, and changes in American class structure.

While all of those factors helped shape popular food trends, it is the latter that offers the best explanation for the rise, fall, and return of the four ideals over the course of the twentieth century. Etiquette guides published in the 1880s, such as the one the epigraph comes from, attest to the importance of dining as an arena of social judgment and suggest that there was an audience of aspirants—people who were unfamiliar with the rituals of fine dining but wanted to learn the "gentle courtesy of manner" that would enable them to imitate people of "high breeding." From slimming to slumming, what the new trends in the Gilded Age have in common is that they were all ways of performing a certain kind of status.

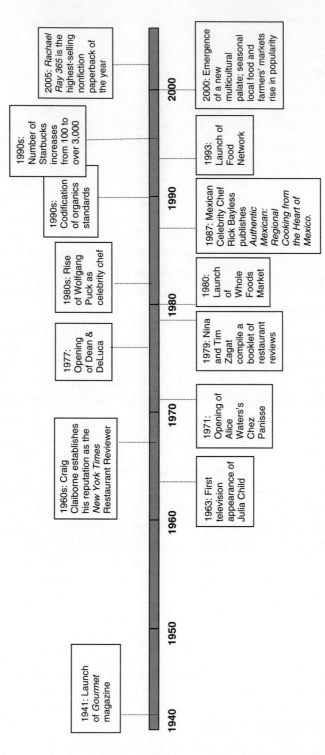

Figure 3. The launch of *Gourmet* magazine in 1941 begins the story of gourmet culture. Source: Josée Johnston and Shyon Baumann, *Foodies: Democracy and Distinction in the Gourmet Foodscape* (New York: Routledge, 2010), 6. Reproduced by permission.

1940   1950   1960   1970   1980   1990   2000

1941: Launch of *Gourmet* magazine

1960s: Craig Claiborne establishes his reputation as the *New York Times* Restaurant Reviewer

1963: First television appearance of Julia Child

1971: Opening of Alice Waters's Chez Panisse

1977: Opening of Dean & DeLuca

1979: Nina and Tim Zagat compile a booklet of restaurant reviews

1980s: Rise of Wolfgang Puck as celebrity chef

1980: Launch of Whole Foods Market

1987: Mexican Celebrity Chef Rick Bayless publishes *Authentic Mexican: Regional Cooking from the Heart of Mexico.*

1990s: Codification of organics standards

1990s: Number of Starbucks increases from 100 to over 3,000

1993: Launch of Food Network

2000: Emergence of a new multicultural palate; seasonal local food and farmers' markets rise in popularity

2005: *Rachael Ray 365* is the highest-selling nonfiction paperback of the year

The Four Ideals in the Gilded Age

The term Gilded Age typically refers to the period from roughly 1870 to 1900 and takes its name from an 1873 novel by Mark Twain and Charles Dudley Warner about people trying to get rich through land speculation. Gilding offered a useful metaphor for the excesses and pretensions of the newly rich, which Twain and Warner portrayed as a thin veneer of luxury and respectability over the crass materialism and corporate graft that drove the accumulation of wealth. Although the novel was neither a critical nor a commercial success, the title was adopted as a convenient shorthand for the idea that America had entered a new era of great fortunes fueled by rampant greed and corruption. Recently, some historians have begun to question the traditionally jaundiced view of the era, noting that many of its most vocal critics stood to gain by exaggerating stories of corruption and that later scholars may have accepted those prejudiced assessments too readily.[3] However, there is little dispute that the dramatic changes that followed the Civil War and Reconstruction concentrated wealth and power in the business-owning and professional-managerial classes of America's growing cities.

Between 1870 and 1900, the country's population nearly doubled largely due to an influx of immigrants, many of whom sought work in the factories concentrated in northern cities. Millions of native-born Americans also moved from the country to the city, including many African Americans who moved north seeking refuge from the terror of the Jim Crow South. The percentage of the workforce directly involved in agriculture declined from around 50 percent in 1870 to 35 percent 1900, but thanks largely to the increasing mechanization of farm work, wheat production increased from 254 to 599 million bushels in the same period. Steel production rose even more dramatically, from 77,000 tons in 1870 to 11.2 million tons in 1900. Some of that steel was used to build railroads that dramatically reduced the time required to travel and transport goods across the country. The telephone, which was invented in 1876, also helped connect the country.[4] Those changes, along with the spread of public sanitation and gas and electric service, raised the overall standard of living for many Americans.

In most other ways, the fruits of America's industrial transformations, including the sixfold increase in the gross national product, were unevenly shared. The transition from self-employment to the wage system was already well under way by 1870, but it intensified in the last three decades of the century. Small manufacturing firms that employed skilled artisans were increasingly replaced by larger, mechanized factories that employed workers who could be replaced more easily. Twelve-hour workdays and poverty-level wages were common. Poor working conditions led to fierce battles between labor unions and managers, but labor claimed few victories. Courts outlawed tactics such as citywide boycotts and declared the eight-hour workday an unconstitutional restriction on the power of employers. State and federal troops were called in to crush strikes. Workers were also vulnerable to market contractions. During the severe economic depressions of 1873–1877 and 1893–1897, unemployment rose to over 16 percent.[5]

Meanwhile, members of the growing professional middle class and some skilled laborers had more leisure time and disposable income than anyone but the very rich had previously enjoyed, creating a demand for mass entertainment and a great proliferation of consumer goods. Minstrel theater, vaudeville, Wild West shows, amusement parks, circuses, and midways attracted a broad and relatively diverse audience, although most theaters barred or segregated African Americans. Civic organizations run mostly by upper-middle-class women also sought to foster high culture, such as art galleries and opera.[6] The women in that demographic were the first to popularize French food in America, diet to lose weight, campaign for pure food, and seek out diverse cuisines.

The Ideal of Sophistication

Americans have associated French food with luxury and refinement since at least the days of the early republic, so its status serves as a proxy for the popular interest in the kind of culinary sophistication designated by the term gourmet. Thomas Jefferson is hailed as America's first gourmet because of the affinity for French food he developed during his diplomatic service. He had sorrel and tarragon planted in his gardens at Monticello. As president, he was known for hosting lavish banquets that featured French wines, labor-intensive delicacies such as ice cream, and *service à la française*, where all the main dishes are placed on the table at once.[7] This style of service had "the advantage of displaying the culinary labor as well as the most variegated and rare products by exhibiting them in all their profuseness," according to the head chef at Delmonico's, the French restaurant regarded as one of New York City's finest in the nineteenth and twentieth centuries.[8] Cookbook author Hilde Lee says Jefferson was responsible for "bringing the European refinements of food to America" and catalyzing a transition from the "mundane meat and potatoes and stews and open-hearth cooking of Colonial times to the beginnings of a more sophisticated cuisine."[9] However, according to historian Harvey Levenstein, very few Americans showed any interest in the culinary refinements associated with continental Europe during the eighteenth and nineteenth centuries. A small elite could afford to eat at restaurants such as Delmonico's, hire French-trained chefs, and drink imported champagne, but those practices "never entered the mainstream."[10]

Instead, French food was an object of populist derision for most of the nineteenth century. During Martin Van Buren's bid for reelection in 1840, the opposition Whig Party used the fact that he had hired a French chef for the White House against him "in a smear campaign labeling him an aristocrat intent on the restoration of monarchy."[11] As late as 1869, housekeeping guides such as *American Women's Home* assumed that their readers would regard "French whim-whams" with disdain.[12] Cookbook authors who dared to include French dishes were defensive about them. For example, in her 1864 book *House and Home Papers*, Harriet Beecher Stowe claimed that she ought to be able to take some leaves from French cookbooks "without accusations of foreign foppery."[13] Like "whim-whams," the word "foppery" implied that French food was fussy and pretentious. Similarly, Juliet Corson's 1877 *Cooking Manual* praised the thriftiness of French home cooks in an apparent attempt to counter the idea that French cooking was inherently fancy and profligate.[14]

The tentative, defensive embrace of French food in Stowe's and Corson's cookbooks turned out to be a bellwether. The idea that it could be thrifty never caught on, but attitudes toward the frills once dismissed as foppery began to shift. According to Levenstein, the first signs of the broader appeal of French food appeared in postbellum menus from upper-class hotels.[15] By the end of the nineteenth century, cookbook writers had stopped being defensive about French influences and instead "assumed that foreignness had cachet."[16] In *Miss Parola's Kitchen Companion*, a popular 1887 housekeeping guide aimed at the middle class, a recommended bill of fare for a dinner for company in autumn or winter consisted of crab bisque, turbot a la crème, potato croquettes, Parisian Vol-au-Vent (roast chicken served with dressed celery and potato puffs), iced orange granité, fried frog's legs with tartar sauce, cauliflower au gratin, reed birds roasted in sweet potatoes served with green peas and lettuce salad, and, for dessert, peaches in jelly with whipped cream, coffee mousse, vanilla ice cream, lemon sherbet, and coffee.[17] As Levenstein notes, "We will never know if anyone ever served exactly that meal, although given the popularity of Parola's books, it is likely that some did. However, what is interesting is that a mere fifteen years earlier it would have been inconceivable for anyone to even suggest that a dinner of this sophistication could or indeed should be cooked and served in a middle-class home."[18]

Another sign of the increasing influence of French culinary style was the spread of *service à la russe*, where courses are served sequentially rather than all at once. This new style of service had become the standard in France by the end of the nineteenth century, and fashionable American restaurants followed suit. In 1894, a former Delmonico's chef noted, "The old style of French service threatens to disappear entirely and is rarely used, except on very rare occasions."[19] *Service à la française* might have provided a more spectacular display of culinary labor, but *service à la russe* actually required more work to pull off. The new style of service contributed to the growing obsession with what women's magazines referred to as "the servant problem"—a shortage of people willing to work in domestic service, particularly those with European training.[20] According to Mary Sherwood, an etiquette writer who contributed regularly to the fashion magazine *Harper's Bazaar*, a dinner party with *service à la russe* required at least one servant for every three guests. However, she also described a modified version that even "young couples with small means" might achieve that put the responsibility for carving the meat in the hands of the host and suggested serving champagne and claret throughout the meal instead of offering a different wine with each course.[21] As with Parola's elaborate bills of fare, we can't know how many people actually followed Sherwood's advice, but the expectation that even a couple of "small means" ought to be able to serve two kinds of imported wine with dinner attests to the growing reach of this aspirational kind of entertaining.

The tide began to turn again at the end of the 1920s. A 1929 article in the trade publication *Factory and Industrial Management* about the Fred Harvey system, a chain of high-end dining cars and train station restaurants on the Santa Fe Railroad, captures both the rise and fall of the French fad. According to the article, Harvey had "upset all western precedent" in 1876 when he hired a French chef from the East Coast to run the new chain. The unconventional move had succeeded in providing passengers with restaurants that "would have been a credit to any large city of the East." However, by 1929, passengers' tastes were beginning to change. A recent experiment done by the superintendent of the dining cars had revealed that "strangely enough . . . if a dish is described with a free use of French, rather than in plain English, the wary diner hesitates to order it; Filet Mignon, Champignons, will not sell in nearly so large a quantity as Small Tenderloin, Mushrooms."[22] Given that the article describes the results of the experiment as strange, it's clear that French food was still seen as desirable, but its appeal—even for the well-heeled patrons of the Fred Harvey system—was apparently on the wane. This shift affected the large cities of the East too. According to Levenstein, "By 1930 the ranks of the French restaurants had been decimated. . . . Moreover, a general prejudice against French cooking seem to have reemerged, alongside widespread ignorance of what it was."[23] Prohibition contributed to the demise of some restaurants by eliminating one of their primary sources of profit, but even after Prohibition was repealed, fine dining was slow to recover. The popularity of foods seen as gourmet gave way to those seen as simple, economical, and wholesome.

The Ideal of Thinness

At the same time that ideal dinner party menus began to grow, the ideal body began to shrink. Most historians agree that dieting specifically to achieve thinness first became widespread in America during the last decades of the nineteenth century, a trend that also began with the urban elite and the professional middle class. Before that, most diet advice was concerned more with spiritual purity or discomforts such as indigestion than with the pursuit of any particular body size. The early nineteenth-century disciples of Presbyterian minister Sylvester Graham followed what sounds like a modern weight-loss diet: it consisted mostly of fruits, vegetables, and whole grains and little or no meat, fat, and alcohol. But Graham and his followers said they wanted to be "robust," not thin.[24] Grahamites weighed themselves and the food they ate to prove they were not losing weight even though they were following such an abstemious diet. Furthermore, even at the height of Graham's popularity, he and his diet were widely mocked, sometimes specifically because it was suspected of causing weight loss despite his assurances to the contrary. According to Hillel Schwartz, both the appearance of thinness and the practice of abstemious eating "bespoke lack of charity and a denial of a rightful American abundance."[25]

American beauty ideals in the early nineteenth century celebrated abundance, especially as represented by the voluptuous female body. According to fashion historian Valerie Steele, for most of the Victorian era, "except for the waist (and the hands and feet) the entire body was supposed to be well-padded with flesh."[26] Corsets could be used to cinch the waist, but nothing could substitute for the desired fullness of the bust and hips. A late nineteenth-century instructional book titled Beauty and How to Keep It claimed that "extreme thinness is a much more cruel enemy to beauty than extreme stoutness."[27] The singer and actress Lillian Russell, who was the highest-paid entertainer in the country at the height of her popularity, was celebrated as much for her voluptuous figure as for her performing talents. At five foot, six inches and 200 pounds (which would be classified as "obese" today), she was hailed as "America's Beauty." Newspapers also reported admiringly on her voracious appetite, especially in her dining exploits with the millionaire "Diamond Jim" Brady. According to one biographer, "Together, they achieved great gourmand triumphs in Chicago restaurants, highlighted by contests to determine who, at one sitting, could eat the most corn-on-the-cob, Brady's favorite food. Reporters covering these contests would bet among themselves about how many ears of corn might be consumed by the contestants. Brady would always win (legend had it that he possessed an enlarged stomach), but Lillian proved fiercely competitive."[28]

By 1894, the worm had turned. That year, a review of a comic opera starring Russell described her beauty as "the doughy sort that ravishes the fancy of the half-baked worshippers of pink-and-white pudginess."[29] After reviews began comparing Russell to a white elephant, she began a very public career in slimming, and "for the next decade or so her diet regimen was as much publicized by the press as any other detail of her life."[30] Russell wrote an occasional advice column for the *Chicago Tribune* that came to focus increasingly on the virtues of bicycling and the dangers of "intemperance in eating," which the formerly competitive eater warned was "almost as fatal to beauty as intemperance in drinking."[31] At the same time, actresses once described as ugly because they were too thin, such as Sarah Bernhardt and Lillie Langtry, became beautiful seemingly overnight.[32]

Like the simultaneous rise of French whim-whams, thinness was initially a preoccupation of the bourgeois. Anorexia nervosa was initially diagnosed in adolescent girls of "high-born families" in the United States and Britain in the 1870s. An older tradition of fasting, sometimes to the point of starvation, existed, especially among religious girls and women, but the new medicalized phenomenon has generally been interpreted as a response to the larger meanings of food in the nineteenth century. Appetite was seen as a proxy for sexual desire and physical robustness, and openly admitting to being hungry or enjoying food was seen as unbecoming in a lady and was associated with the working classes. Joan Jacobs Brumberg notes that "historical evidence suggests that many women managed their food and their appetite in response to the notion that sturdiness in women implied low status, a lack of gentility, and even vulgarity. Eating less rather than more became a preferred pattern for those who were status conscious."[33] It's notable that when adolescent girls first began deliberately trying to get thin, thinness was seen as a sign of sickness, not health, and that that was actually part of its appeal. Like the eighteenth-century Parisians who frequented the first restaurants, which were named for the "restorative" broths they served,[34] the fasting girls of the late nineteenth century were deliberately cultivating an aura of ill health that at the time signified privilege.

In the last decades of the nineteenth century, the new ideal of thinness spread through the middle classes. Brumberg says that "by 1900 the imperative to be thin was pervasive."[35] Illustrations and advertisements in national magazines popularized the image of a "new American woman" who was significantly slimmer and more active than her voluptuous Victorian era counterparts, exemplified by the illustrations of Charles Dana Gibson.[36] The Gibson girl gave way to the even thinner flapper, represented by early film stars such as Mary Pickford and Clara Bow, who were frequently described as "small and boyish."[37] Instead of a curvaceous body with an exaggerated (often corseted) waist, the ideal flapper had virtually no curves, like the heroine of Aldous Huxley's novel *Antic Hay*, whom he described as "flexible and tubular, like a section of a boa constrictor . . . dressed in clothes that emphasized her serpentine slimness."[38] Masculine bodily ideals changed too. In an example that marks the ambivalent transitional period, Schwartz notes that during Grover Cleveland's 1882 New York gubernatorial campaign, his supporters proudly advertised their candidate's large frame and inclination to corpulence as signs of gravitas while his opponents portrayed his 250 pounds as a sign of weakness. Although Cleveland won the election, the idea that fatness in men might be admirable was on its way out.

Laura Fraser's history of the diet industry also identifies 1880–1920 as the crucial period when fatness morphed from a sign of health and beauty to a sign of disease, ugliness, and moral weakness in America. The shift is apparent in the magazine columns written by a medical school professor named Woods Hutchinson. In 1894, he responded to what was apparently still a nascent demonization of fat by attempting to reassure the readers of *Cosmopolitan* magazine that fat was both medically benign and attractive. He described fat as "a most harmless, healthful, innocent tissue" and claimed that "if a poll of beautiful women were taken in any city, there would be at least three times as many plump ones as slender ones."[39] He also warned against starving or exercise as means of weight control, practices that Fraser claimed were then "just becoming popular."[40] It's clear that by 1926, the forces aligned against fatness had triumphed. In an article for the *Saturday Evening Post*, Hutchinson wrote, "In this present onslaught upon one of the most peaceable, useful and law-abiding of all our tissues, fashion has apparently the backing of grave physicians, of food reformers and physical trainers, and even of great insurance companies, all chanting in unison the new commandment of fashion: 'Thou shalt be thin!'"[41] Hutchinson wasn't the only holdout. Another physician Schwartz quoted wrote in 1927: "Concerning obesity, the amount of scientific information which we have regarding it is in marked contrast to the large amount of public opinion on this subject."[42] Nonetheless, as Hutchinson complained, many other doctors were willing to lend medical credibility to the new fashion.

As thinness became fashionable, many diets and medicines that had previously been touted as cures for dyspepsia or nervous exhaustion were recuperated as slimming regimens. Additionally, several new techniques emerged that were specifically designed to promote weight loss: fasting, Fletcherism, calorie counting, and thyroid medication.[43] All of these techniques were either inspired or given post hoc justification by the new nutritional logic ushered in by the calorie, which was introduced to the American public by the chemist W. O. Atwater in a series of articles in *Century* magazine.[44]

Just as we can't know how many people participated in multicourse dinner parties with any regularity, it's impossible to know exactly how many people actually engaged in dieting to lose weight at the turn of the last century or who exactly was taking part in the new trend. A 1925 article in American magazine claimed that "reducing has become a national pastime . . . a craze, a national fanaticism, a frenzy."[45] However, then, as now, the practices of the urban middle classes were far more likely to be recognized as universal than whatever poor, nonwhite, and rural Americans were doing. Although most diet techniques were popularized by men, the paradigmatic weight-loss dieter was a woman.[46]

It's also difficult to say exactly when slimming began to fall out of fashion, but several accounts locate a shift in attitudes toward thinness around the time of the Great Depression. In 1931, the director of the Carnegie Institute's Nutritional Laboratory, which put on lectures for the public, wrote that the frenzied interest in weight reduction had "finally receded a bit."[47] In his critique of obesity politics, J. Eric Oliver claims the "idealization of thinness was largely suspended during the privations of the Great Depression and World War II."[48] In place of the streamlined flapper, the voluptuous hourglass figure returned, epitomized in the mid-twentieth century by actresses such as Marilyn Monroe, Sophia Loren, and Elizabeth Taylor.[49] In place of the universal logic of calorie math, many Americans embraced the idea that people were born with a particular build that was resistant to attempts to change it. According to William Sheldon's theory of somatotypes, which was popularized by a 1940 *Time* magazine cover article, those builds were based on what kind of embryonic tissue had dominated in the womb—a larger ectoderm would make someone a slender ectomorph, a dominant mesoderm would produce a muscular mesomorph, and a substantial endoderm would result in a plump endomorph. Somatotype theory suggested that instead of being infinitely malleable, body shape was powerfully influenced by factors beyond individual control.[50] Throughout those decades, new diet programs and products such as Weight Watchers and amphetamine pills continued to appear, but dieting to lose weight generally came to be seen as a fringe and possibly futile pursuit for people with "weight problems." The sudden resurgence of mass dieting and exercise for weight loss in the 1980s was hailed as a new national obsession, as if the Gilded Age fad for reducing had never happened.

The Ideal of Purity

The pure food movement has survived in popular memory better than Gilded Age dinner parties and slimming techniques such as Fletcherism thanks in part to *The Jungle*, the book that made Upton Sinclair famous before he became a fasting guru. The movement also left a bureaucratic legacy in the form of the FDA, which grew out of the regulatory powers established by the Pure Food and Drug Act of 1906. However, there are some significant differences between the standard account of the movement as told by high school history textbooks and some contemporary food writers and what most historians have to say about it. The standard account tends to credit *The Jungle*, which became a best seller within weeks of its publication in 1906, with catalyzing the movement. In *Fast Food Nation*, Eric Schlosser claims that "the popular outrage inspired by *The Jungle* led Congress to enact food safety legislation in 1906."[51] Most U.S. history textbooks emphasize the role played by Teddy Roosevelt but similarly place the key events in the lead-up to the Pure Food and Drug Act after the turn of the century. For example, the 2007 edition of Gerald A. Danzer and colleagues' *The Americans* says, "After reading *The Jungle* by Upton Sinclair, Roosevelt responded to the public's clamor for action. He appointed a commission of experts to investigate the meatpacking industry. The commission issued a scathing report backing up Sinclair's account of the disgusting conditions in the industry. True to his word, in 1906 Roosevelt pushed for passage of the Meat Inspection Act."[52] Most historians, on the other hand, date the rise of the pure food movement earlier and argue that instead of being prompted by muckraking journalists and reform-minded politicians in the Progressive Era, it was led by urban, upper-middle-class women.

According to Lorine Swainston Goodwin's history of the pure food and drug movement, grassroots campaigns that protested the sale of adulterated or mislabeled food, urged their fellow citizens to practice temperance and eat more whole grains, and demanded more government regulation in the food industry began to emerge in the late 1870s and early 1880s. Mary Hackett Hunt's campaign against the abuse of drugs and alcohol began with a town hall meeting in Massachusetts on Easter Sunday in 1879. Ella Kellogg founded the Health and Heredity Normal Institute in Battle Creek, Michigan, which sought to teach women how to avoid "diseased" foods such as refined sugar, white flour, and spicy or pungent condiments, in January 1884. The Ladies'

Health Protective Association was formed in December 1884 by fifteen women who lived in the exclusive Manhattan neighborhood of Beekman Place and sought to pressure the city to make a nearby slaughterhouse and fertilizer dealer clean up its act. Soon, all across the country, long before *The Jungle* was published, women were organizing to press for pure food.[53]

There were significant differences between the goals and methods of these different campaigns. While Hunt was advocating the legal prohibition of alcohol, other pure food activists complained that "first class barrooms" sold an "adulterated article" instead of real beer or whiskey and urged Congress to pass laws to give the country "pure licker." Kellogg's institute advocated a vegetarian diet, arguing that "man is not naturally a flesh-eater," whereas the Ladies' Health Protective Association sought to ensure that the meat they intended to keep feeding their families was clean and safe (and, ideally, butchered somewhere out of sight). Hunt and Kellogg largely targeted individual consumers, whereas the Ladies' Health Protective Association went directly to regulatory agencies and the elected officials that oversaw them. There were also always multiple forms of purity in play—the purity of the body that was free from substances that were seen as unnatural, overly stimulating, or enslaving; the purity of food free of contaminants; and the purity of the factories and industries that produced food and drugs.[54]

However, despite the differences among them, these efforts were seen as part of a relatively coherent movement, even in those early days. Thus, Goodwin claims, "Dating the beginning of the fight for pure food and drugs after the turn of the century is clearly too late. . . . A preponderance of the evidence reveals that the demand to improve the quality of food, drink, and drugs gained momentum during the three preceding decades."[55] She suggests that some people have been misled by the fact that "politicians, bureaucrats, and journalists of the early 1900s often wrote of food and drug adulteration as if it was their exclusive discovery."[56] After the 1906 act passed, they rushed to take credit for its passage. But according to Goodwin, crediting people such as Sinclair and Roosevelt (or USDA chemist Harvey Wiley, sometimes called the "Father of the Pure Food and Drug Act") gets things precisely backward. *The Jungle* didn't catalyze the fear and outrage about adulterated food. It fed on the existing groundswell of outrage cultivated by decades of grassroots activism. At best, *The Jungle* fanned the flames, but it might be more accurate to see it as merely riding on the coattails (or perhaps the petticoats) of organized women.

The legislative record backs Goodwin up. It's certainly possible that the sudden success of *The Jungle* helped the 1906 act pass the House of Representatives. However, the act passed the Senate on February 12, 1906, with only four votes against, weeks before Sinclair's novel ever reached bookstores.[57] Furthermore, both branches of Congress had passed similar bills before, beginning with Nebraska senator A. S. Paddock's sweeping food and drug bill of 1889. None of the earlier bills had managed to clear both houses of Congress, but the success of the 1906 act probably had more to do with a compromise it offered on the funding of inspections than with Sinclair's book. Whereas many of the previous bills would have made industry pay for inspections, the act that finally passed called for federal funding. This also suggests that the high school textbook narrative portraying the act as one of Teddy Roosevelt's triumphs in the battle against big business is not wholly accurate either. In fact, many industry giants such as the Pabst brewing company, Old Taylor whiskey, and the H. J. Heinz company lobbied in support of federal regulation, particularly if it could be done on the government's dime. Brewers and distillers hoped regulation might stave off full-scale prohibition, and producers such as Heinz predicted (accurately) that regulations would help eliminate competition from smaller businesses unable to meet the new standards.[58]

Goodwin's theory that the existing concern about impure food catapulted The Jungle to success, which contrasts with the narrative that claims that it was the book that prompted the outrage, also helps explain why the book's reception and legacy are so at odds with its content. Although it is often portrayed as a journalistic exposé of the meatpacking industry, it's actually work of fiction. The main characters sometimes do work in parts of the meatpacking industry, including the stockyards, a fertilizer plant, a lard cannery, and a sausage-making factory, but they also find work in brothels, a can-painting shop, a pickle factory, a harvester factory, a steel mill, and a hotel. They take odd jobs cutting blocks of ice, doing farm work, and helping rig an election. The descriptions of unsafe and unsanitary conditions in meatpacking that became the focus of the public reaction to the book occupy scarcely a dozen of the novel's 400 pages. Sinclair himself was famously disappointed in the book's impact, later writing, "I aimed for the public's heart and by accident hit it in the stomach." In Eric Schlosser's foreword to the 2006 Penguin edition, he interprets this to mean it had been written "to help meatpacking workers, not to improve the quality of meat,"[59] but it would be more accurate to say it was written to protest the "ferocious barbarities of capitalism,"[60] not just for their impact on meatpacking workers but for the whole downtrodden working class.

The activists responsible for cultivating the public outrage about impure food that caused people to interpret The Jungle as a book about unclean slaughterhouses were primarily professional middle class women.[61] While most women's clubs were formally open to all women, the leadership was primarily drawn from a narrow demographic. Sarah Sophia Chase Harris Platt-Decker of Denver, Colorado, the first president of the GFWC, was married first to a business tycoon who was active in politics and then to Denver judge W. S. Decker. Eva Perry Moore, the first vice-president, was a Vassar graduate who had traveled in Europe between graduation and her marriage to an engineer who became the president of several iron mines. Helen Miller, the first chair of the GFWC's Pure Food Subcommittee, was a home economist at the Agricultural College in Columbia Missouri and was married to another academic. Additionally, the fact that the authors of the General Federation of Women's Clubs (GFWC) official charter felt it necessary to stipulate that membership was open to any woman regardless of her "station in life" says plenty about the exclusionary tendencies of some member clubs.

The status of the leadership was sometimes reflected in the broader membership and club practices. For example, Decker once visited a club in Denver that met in a "wealthy and handsome clubhouse" in a room with a decorative inlaid border on the floor. The president of the club told her that there were 200 members and as many more on a waiting list. Decker asked why the membership was restricted and was told that only 200 chairs would fit within the border. Setting chairs on the border itself was not to be considered.[62] Another sign of the clubs' functional exclusivity is the parallel rise of clubs formed by and for African American women. Some women belonged to both, like Josephine St. Pierre Ruffin, a biracial woman from Boston who became one of the early leaders of the National Association of Colored Women and was also a co-founder of the majority white New England Women's Club and American Woman Suffrage Association. However, African American women's clubs were also largely a middle-class phenomenon and typically focused more on anti-lynching campaigns, suffrage, and black women's uplift than on food purity.[63]

The activism of professional middle class white women related to pure food did not stop with the passage of the 1906 act. On the contrary, the movement was energized by the victory. According to Goodwin, "an extraordinary explosion of activism accompanied attempts to implement the law" as the same organizations that had pushed for its passage organized around the conviction that "the Augean stables are still unclean."[64] She locates the beginning of the decline in 1912, as some of the old leadership began to leave the clubs or divert more of their energies into the campaign for Prohibition. Other histories of the clubwomen's movement place the decline a little later, usually at the end of the 1920s.[65] The consensus is that, much like fancy dinner parties and slimming diets, grassroots organizing around issues of food purity subsided almost entirely during and after the Great Depression.

The Ideal of Cosmopolitanism

The last major shift in the culinary trends of the Gilded Age was a growing interest in international and ethnic foods. The sudden popularity of French food may also have been part of this new culinary cosmopolitanism, but the trend encompassed many cuisines that were not generally seen as sophisticated. Foreignness itself developed a sudden cachet. This marked a significant departure from the staunch culinary nationalism that had shaped most American food writing through the 1860s.[66] As late as 1872, a *New York Times* editorial lamented that although the city's population was growing more diverse, its cuisine was not: "New York, we all say, is getting to be very cosmopolitan . . . [and] it might be supposed that eclectic modes of cookery would spring up. . . . We may be ever so cosmopolitan, progressive, and modern in other things, but in [our cookery] we are still far behind hand." Although cosmopolitan cookery might not yet have caught on with the general population, the editorial's suggestion that it was desirable was a portent of things to come.[67]

Once again, the shift seems to have happened in the 1880s. In March 1880, an article in the *New York Tribune* mentioned "the growing interest and even enthusiasm of housekeepers in all parts of the country in acquiring the secrets of foreign cooks and learning their manner of preparing the economical, wholesome, and appetizing dishes of which Americans have been so curiously ignorant."[68] In the last two decades of the nineteenth century, cookbooks and women's magazines began publishing recipes that advertised their foreignness. In 1888, *Good Housekeeping* reported that national suppers, meals that might feature either a smorgasbord of dishes seen as representative of different countries or a menu based on a single national theme, were "much in vogue."[69] By 1902, even modestly titled books such as *Practical Cooking* by Janet McKenzie Hill featured recipes for "Asparagus, Spanish Style" and "Zwieback (Berliner Frau)." Hill also reassured hostesses that they need not serve "Flamingoes from Sweden, game from Africa and South America, and pears from Assyria" all in the same meal to impress their guests, suggesting that her readers would not only assume foreignness was fashionable, they might even be feeling overwhelmed by the pressure to wow people with exotic fare.

Food manufacturers also began catering to more cosmopolitan tastes. By 1905, Campbell's canned soup varieties included consommé, julienne, chicken gumbo, pepper pot, vermicelli-tomato, and mulligatawny.[70] Armour and Company and Libby, McNeill and Libby offered canned chili con carne and tamales. According to Kristin Hoganson's history of postbellum consumer culture, "food writings aimed at middle-class American women provided instructions for everything from English peas to Flemish fish, Hungarian goulash, Russian Piroga, Dutch Pudding, Venetian fritters, Vienna bread, German liver dumplings, Spanish olla podrida, Swiss eggs, and oatmeal stuffed haddock as eaten in Aberdeen and Limerick."[71] Although the majority of the references were to European cuisines, some food writing featured recipes from other parts of the world. The "Foreign Cooking" section of a 1918 housekeeping manual by Lucia Allen Baxter included recipes for Turkish kabobs, Mexican frijoles, East Indian sholah pullow, Cuban corn pie, Bengal chutney, and English tea cake from the Bermudas. A 1910 *Good Housekeeping* article provided instructions for producing a Mexican dinner that included chili bisque, *tomate con queso*, *chilies relleno*, and *cidracayote* (squash and corn).[72] Even if no one ever cooked that menu, it's clear that food writers at the turn of the century believed that their middle-class audience at the very least wanted to read about foreign foods.

Both private citizens and civic organizations hosted what they called "foreign entertainments." In the invitations to these events, guests were sometimes encouraged to come in costume, building on the older trends of masked balls and parlor theatricality.[73] Japanese teas were particularly popular. *Demorest's Family Magazine* described a Japanese tea in 1892 "given by a family who have made the fashionable trip to Japan and brought back a choice variety of Japanese costumes and bric-a-brac." The report continued, "They received their friends in a delightfully informal manner, each of the hostesses wearing a veritable Japanese dress, of crepe, magnificently embroidered, and their heads bedecked with pins galore. Each had her tiny table from which she dispensed delicious tea in rare Japanese cups, both the cups and their contents eliciting enthusiastic exclamations from the recipients."[74] Women's clubs, both white and African American, held Japanese teas to raise money for charitable causes.[75] These were so common that Gilded Age entertaining guru Ellye Howell Glover wrote in her 1912 *Dame Curtsey's Book of Party Pastimes for the Up-to-Date Hostess*: "While there is nothing very new in the Japanese scheme, it is always effective and people never seem to tire of it."[76]

This was also the period when native-born Americans, especially in the professional middle class, began patronizing ethnic restaurants in greater numbers. Until at least 1880, most ethnic restaurants catered primarily, if not exclusively, to immigrants from their own countries.[77] An 1868 article on San Francisco's restaurant scene in *Overland Monthly* reported that "Germany has several restaurants—not especially distinctive, but essentially Germanesque in their customers. . . . The Italian restaurants, however, are more exclusively patronized by the people of their own nationality than is true of any other class." Of the restaurants in the city's burgeoning Chinatown, the author wrote: "Few western palates can endure even the most delicate of their dishes."[78]

By the 1890s, it was assumed that middle-class diners would want to explore the best that every nationality had to offer. An 1892 article in *Overland Monthly* claimed that "so far as the education of the stomach goes, one may obtain all the benefits of the grand tour without leaving San Francisco." The author recommended the herring salad and matzos at a German-Jewish restaurant, a Mexican restaurant "where the visitor can burn out his alimentary canal in the most approved Spanish style," a "genuine" Italian restaurant, and a restaurant in Chinatown with "a wonderful assortment of viands, even to shark's fins and birdsnest soup."[79] Japanese noodle houses serving udon and ramen catered to the working class and in places such as Ogden, Utah, were known as sites of gambling, fighting, and frequent police raids.[80] Foreign foods were no longer just for the foreign born. The owners and staff of many ethnic restaurants played up their foreignness for crossover consumers with exaggerated costumes and folk music, performing a sort of nationality drag in order to create a more exotic dining experience.[81]

In her history of ethnic food in America, Donna Gabbacia describes the first decades of the twentieth century as a particularly "intensive phase of cross-cultural borrowing." After the 1920s, people continued to eat foods once associated with immigrants, but the way people thought about those foods had changed. Because of the privations of the First World War and the Great Depression and the gradual assimilation of the immigrant populations they were associated with, dishes such as spaghetti and chop suey were increasingly portrayed as familiar and patriotic, not exotic or novel.[82] Foods that still seemed foreign to American eaters lost their popular appeal, and food manufacturers adjusted accordingly. For example, Campbell's mulligatawny soup was discontinued in the mid-1930s.[83] By the Second World War, ethnic foods were celebrated for being cheap and requiring less meat than many dishes seen as classically American. Gabaccia welcomes this as "culinary pluralism," but it was no longer culinary cosmopolitanism.

The general retrenchment from culinary adventurousness to a preference for the familiar and plain persisted until 1980. Even in New York City, a journalist who surveyed 200 Manhattan shoppers in 1961 concluded that "most Manhattanites eat well . . . if unimaginatively."[84] Aside from a few exceptions, "when asked for a typical dinner menu, a majority of those asked listed steaks or chops, a green or yellow vegetable, salad and gelatin or canned or fresh fruit for dessert. In moderate to upper income homes, pot roast, beef stew, roast lamb, pork or beef, and roasted chicken are some other popular meat dishes."[85] Only a bohemian couple interviewed in Greenwich Village exhibited cosmopolitan tastes, saying that they enjoyed the French, Italian, and German foods they had come to know and like when they traveled in Europe; Arabic dishes prepared by the wife, who had been born in Egypt; and the "chopped liver, borscht, and gefuellte [sic] fish" the husband had grown up eating. The mainstream palate was reflected better by a woman from Stuyvesant Town, which the article described as populated by "predominantly young, middle-income families," who told the reporter, "My husband likes plain food. . . . Give him steak and he's happy. If I try a new dish, he looks as if he thought I was trying to poison him."[86] As with the other Gilded Age ideals, culinary cosmopolitanism declined to the point that the revival of ethnic and international foods in the 1980s, especially among young urban professionals, could be hailed as something remarkable and new.

Alternative Explanations

Although many historians agree there were major shifts in American food culture around 1880 and again around 1930, they don't always agree about the causes of those shifts. Like the food revolution that started in the 1980s, the Gilded Age rise of the four ideals is sometimes attributed to progress and enlightenment. However, their decline in the 1930s poses a problem for that logic. If slimming diets and sanitariums became popular because of advances in nutritional science, it is difficult to explain why they would have become less popular in the 1930s. Furthermore, diet fads largely preceded the scientific rationales that were later used to justify them. Similarly, although increasing travel and trade certainly facilitated interest in French cooking and other international foods, access to information and goods alone cannot explain the popular appeal of the exotic. The immigrant communities that provided cosmopolitan diners at the turn of the century with a growing array of novel foods didn't simply evaporate in the 1930s, when plain home cooking came to the fore. Many of the most successful brands that were launched during and after the world wars were started by immigrants, including Swanson's TV dinners and Chef Boyardee (an Americanization of Boiardi), but as Gabbacia notes, their mid-century success came from "selling their products with no ethnic labels attached."[87]

Of the many other theories about why the central ideals of the food revolution first emerged during the Gilded Age, four merit additional consideration because they are occasionally presented as a kind of common sense: 1) the triumph of abundance over scarcity; 2) the rise of new scientific ways of thinking; 3) the growing empowerment of women; and 4) the evils of industrialization.

The emergence of the ideal of thinness in the United States and other Western industrialized nations, in contrast to the globally and historically more prevalent preference for plumpness,[88] has been widely attributed to the increasing availability of food. According to the basic logic of this theory, when food is scarce and expensive, the poor will generally be thin and only the rich can get fat. Thus, fatness will be an emblem of status and usually of health and beauty, too. When food is plentiful and cheap, anyone can get fat, and it may take additional resources (e.g., time to exercise or enough money to buy more expensive food) to get or remain thin, so the status symbols get reversed. Fraser explains the emergence of weight-loss dieting in the 1880s by saying, "When it became possible for people of modest means to become plump, being fat no longer was seen as a sign of prestige. . . . The status symbols flipped: it became chic to be thin and all too ordinary to be overweight."[89] There are two main problems with this theory: food wasn't scarce in America before the Gilded Age, and people did not get significantly fatter during the years when the status symbols flipped.

There was good reason for America's reputation as a "land of plenty" in the sixteenth and seventeenth centuries. According to James McWilliams's history of food in colonial America, the colonies produced more than enough food for its growing population from 1668 to 1772. Describing "unfettered access to food," McWilliams writes, "the average free colonists by the middle of the eighteenth century enjoyed something that their ancestors could only have dreamed of enjoying: the ability to eat and drink more or less what they wanted when they wanted it."[90] European visitors to the United States often commented on the abundance of food, especially meat, that was available to even the poorest Americans. According to Roger Horowitz, the annual meat allowance for widows specified in wills rose from 120 pounds in the early 1700s to over 200 pounds in the early 1800s. The standard annual meat allocation for slaves in the South in the 1800s averaged around 150 pounds per person. Livestock production statistics from that period also "give us some confidence in suggesting an average annual consumption of 150–200 pounds per person in the nineteenth century." That figure is only slightly less than the average today and included larger proportions of beef and pork.[91] Meanwhile, despite tremendous increases in food production during the Gilded Age, there were still widespread reports of hunger in that period, particularly among working-class immigrants. Then, as now, the poor often experienced scarcity even in the land of plenty.

The second problem with the theory of abundance is that there's little evidence that enough people were getting fatter in the Gilded Age and Progressive Era that thinness became rarer and more distinctive. As late as

1926, the American Medical Association admitted to having no reliable standards for the average weight of adults of any age or height. However, people were routinely weighed upon military conscription, at physical exams, and on scales that traveled with public fairs and exhibitions. The available data shows little change in the average height-to-weight ratio from the early nineteenth century to the twentieth century. According to the 1913 *American Physical Education in Review*, on average, "American college students . . . had grown 1 inch and 3 lbs in fifty years."[92] Based on military conscription data, by World War II the average white soldier in his twenties was on average one inch taller and half a pound heavier than the average Union soldier in his twenties who fought in the Civil War. Adults of all ages stepped on the scales at state fairs and world expos, and similarly show little change from the 1850s to the 1930s. Schwartz reports, "At New England fairs in 1859, men of all ages taken together had averaged 146 to 152 lbs, women of all ages 124 to 126 lbs; at a fair in Philadelphia in 1875, nearly sixteen thousand men yielded an average weight of 150 lbs while more than seventeen thousand women stood at 129 lbs; at the Century of Progress Exposition in Chicago in 1933–34, men from twenty-five to sixty ranged in weight from 153 to 171 lbs, while women ranged from 123 to 136½ lbs. Both sexes had gained perhaps two inches and put on perhaps 10 lbs in seventy years."[93] Of course, it's possible that the average weight might have remained roughly constant if poorer people were getting fatter and rich people thinner, although that seems unlikely given that the average weight of incoming college students, a relatively affluent group, increased slightly rather than declining.

A related theory suggests that the ideal of thinness arose not because of any actual increase in the availability of food or the size of most bodies but instead because of a more general sense of excess. This is the theory Schwartz favors; he attributes the new ideal of thinness to "a shift in economic models from scarcity to abundance."[94] Instead of responding to actual physical fatness, he argues, people were responding to the nation's growing wealth. He connects the new fretting about "overnutrition" to Veblen's critique of conspicuous consumption in the leisure class and the idea that fatness represented an outmoded hoarding. Weight-loss dieting was a way to be unburdened, exuberant, and modern.[95] T. J. Jackson Lears makes a similar argument in his history of American advertising, suggesting that the rise of weight-loss dieting was part of a broader "perfectionist project" that developed in response to the new anxieties and desires generated by a culture of excess. Lears contrasts Victorian-era advertisements, in which cornucopias of goods and voluptuous women represented a fantasy of plenty in an age of scarcity, with the slender Gibson girls and reducing diets of the turn of the century. He argues that they represented the new fetishism of efficiency in an age of plenty: "Fat was not a bank account but a burden. What was to be stored was not embodied in flesh but was a more evanescent quality: sheer energy, the capacity for intense, sustained activity."[96] While plausible, the theory of slimming as a reaction to symbolic abundance doesn't explain the reemergence of the ideal of voluptuousness and the decline of weight-loss dieting in the mid-twentieth century. The idealization of buxom beauties and the belief that some people were simply destined to be "husky" no matter how they ate persisted through the country's recovery from the Great Depression and the decades of rapid economic expansion and growing purchasing power for the working and middle classes—an era of plenty and abundance if there ever was one.

Both Lears and Schwartz also connect the growing preference for slimness to the new ethos of scientific rationality that emerged during the Gilded Age. Lears argues that the new, thin iconography of the body was part of a broader transformation in American society that involved an increasing reverence for science. Movements such as Taylorism and home economics endeavored to apply the principles of scientific management to everything from the factory to the office to the home to the body. He connects the rise of weight-loss dieting with a wave of advertisements for intestinal cleansing, soap, and deodorants in the early twentieth century, but according to Lears, "The most immediate and obvious impact of this mechanistic style of thought was a slimming down of body types." According to this theory, fatness was associated with irrational, old-fashioned hoarding whereas thinness was better aligned with the new, streamlined, modern approach.

The calorie and the new theory of energy balance also appealed to the new rational ethos. According to Schwartz, "The calorie promised precision and essence in the same breath. It should have been as easy to put the body in order as it was to put the books in order for a factory."[97] In fact, one of the early cheerleaders for calorie counting was the Yale economist Irving Fischer. A great fan of Taylorism, Fisher found a parallel between Atwater's explanation of energy balance and his own theory of market equilibrium. Attempting to balance calories consumed and calories burned offered a way to apply the same universal, rational logic to the body that was thought to govern the laws of physics and the market. He published a series of articles on slimming through careful calorie accounting and then a best-selling book co-authored with Dr. Eugene Lyman Fisk, *How to Live*, which provided tables of foods containing 100 calories and diet tips such as "Constant vigilance is necessary. . . . Nature counts every calory [*sic*] carefully."[98]

Energy balance still has a certain rational appeal, but that doesn't actually explain the sudden preference for thinness and the popularity of weight-loss

dieting at the turn of the twentieth century. The calorie is simply a measurement of energy. It could have been used to justify diets designed to maintain
a consistent or "robust" weight, such as Sylvester Graham's. The idea that
mechanistic thinking would inevitably lead to a preference for slimmer bodies seems to presuppose that thinness is inherently more efficient, modern,
and rational. Just a few decades earlier, Victorian adolescents had fasted
because thinness was associated with indolence and physical delicacy, not
energy and health. Since the slender body ideal became fashionable before it
was associated with efficiency and modernity, it seems likely that the causal
arrow points the other way: once people began to think about thinness and
weight loss as desirable, they embraced justifications that reflected the new
fetishism of efficiency and rationality.

A third explanation for the new trends looks to the gender ideology of
the era. Women were the primary audience for much of the burgeoning
literature on entertaining and dieting and were the driving force behind
the pure food movement. According to the "cult of domesticity" or "cult of
true womanhood," women's primary sphere of influence was the home. The
pure food movement has been interpreted, along with the many other urban
social reform movements women participated in, as a training ground for
suffragettes and a proxy form of political activism that gave women a voice in
the public sphere before they won the right to vote. According to this theory,
women reformers sought to both work within and expand the domestic
sphere by pursing political change through their accepted role as managers
of the home, particularly the kitchen. Because they focused on issues such
as temperance and pure food, women's clubs were typically seen as outside
or above politics, even when they were pushing for legislative reforms.[99]
According to Karen Blair, women's societies' overt acceptance of traditional
roles gave women an "ideological cover" for their more radical agendas and
for the role of their organizations in carving a larger role for them in the
public sphere.[100]

Similarly, Hoganson argues that cosmopolitan dining gave formally disenfranchised women a way to express their privilege by exploring the bounty
of a growing American empire without actually leaving the domestic sphere.
She argues that women had a particular incentive to acquire foreign goods in
the era before suffrage:

> Although the women I write about exercised social, cultural, and economic
> power over others, they were not political leaders or, in most cases, fully
> enfranchised citizens. Even as they seized the opportunities that were avail-
> able to them, they had grounds to doubt whether they were in control of

their own destinies. This explains much of the attraction of the consumer's imperium. The pleasure of boundless consumption deflected attention from the inequities encountered on the home front by reminding these women that, on a global scale of things, they occupied a position of privilege. The women who bought into the consumer's imperium sought not only tangible items but also a sense of empowerment.[101]

Women found that sense of empowerment, she argues, not just in novel recipes and imported ingredients but also in using their kitchens as places of global encounter where they could learn about the wider world and add adventure to their daily lives. Putting on ethnic costumes and eating foreign foods enabled them to assert their essential whiteness and Americanness in contrast.

The idea that women used food as a kind of nascent feminism might explain why both the pure foods movement and cosmopolitan dining became less popular after the Nineteenth Amendment gave women a formal political voice, although there was some lag between women's suffrage and the decline of both trends. However, this theory doesn't explain the simultaneous rise of Progressive reform sentiment among urban professional men or the fact that men too ate at fancy French restaurants and were the primary target of books on gastronomy. Gender politics undoubtedly shaped both activism and consumption choices, but ultimately the participants in Progressive Era social movements were unified less by gender than by class. The feminization of all the Gilded Age food trends was likely due to the feminization of cooking, shopping, and entertaining and the gendering of beauty ideals rather than to any special use of food by women as a proxy for public sphere engagement.

The final theory, and the one most prevalent in today's popular literature on food, is that the Gilded Age trends were a response to a real decline in food quality caused primarily by the growth of industrial processing. The activists themselves claimed that food adulteration, unsafe medications, alcoholism, and diseases caused by poor diets were on the rise and that these were real and urgent social problems that demanded action. In contrast to the idea that the pure food movement was primarily a way for women to have a voice and only incidentally about food, advocates of the industrialization theory urge us to take the pure food activists at their word about the state of America's food system. But Gilded Age activists typically did not blame industrialization; instead, they complained about unscrupulous corporations, immigrants, and general moral decline. Since blaming immigrants is no longer in fashion, people who support this theory tend to focus on the decline of imagined safeguards such as consumers being able to see for themselves how clean their neighborhood butcher shop was or knowing the farmer who grew their vegetables. The main problem with this theory is that it doesn't fit with the actual growth of the industrial food system or what we know about rates of alcoholism and food adulteration.

According to Richard Hofstadter's work on Progressive Era social reform, the transition from independent yeoman farmers to cash crop farmers dependent on the country store for their supplies was "complete in Ohio by about 1830 and twenty years later in Indiana, Illinois, and Michigan. . . . In so far as this process was unfinished in 1860, the demands of the Civil War brought it to completion."[102] In the industrializing cities of the North, where most pure food activism took place, consumers had been reliant on food that was transported, handled, packaged, and marketed by a series of anonymous middlemen for the better part of a century before organized women and muckraking journalists took up the cause.

Turn-of-the-century cities were dirty and disease ridden, but conditions in the rural areas many people had recently moved from weren't any better. Rural areas were plagued by contagion and sanitation problems too, often caused by the contamination of drinking water and clean soil with animal manure and the vulnerability of large stores of food to vermin. A 1909 handbook on rural hygiene notes that deaths from infectious diseases popularly associated with cities, such as influenza and dysentery, were actually higher in rural districts than urban centers.[103] Data such as that have caused some historians to question pure food reformers' claims that impure food and drink were new or newly pressing concerns in the 1880s. Good evidence about the actual incidence of food poisoning, vitamin deficiencies, and alcoholism is hard to come by and likely wasn't available to the pure food activists either. As far as they were concerned, the rising incidence of these problems was simply obvious. However, many historians argue that contaminated or adulterated food, alcohol use, and dangerous or ineffective drugs had always been social problems.[104] What changed in the late nineteenth century wasn't the extent of the problems but rather the degree of popular concern about them.

It's possible that intensified industrialization may have caused increased anxieties about the food supply even if it didn't worsen problems such as adulteration or alcoholism. But that doesn't explain the dramatic decline in food activism in the 1930s. As many contemporary critics of the food system lament, the continued growth and consolidation of industrial food processing during and after World War II created even more distance between farmers and consumers. It also introduced a growing list of refined and synthetic ingredients into the food supply and created even more opportunities for large-scale contamination. But instead of rising up against a food system that was growing ever larger and more opaque to the common consumer, mid-century Americans embraced the growing array of industrially processed foods to an extent that made the 1960s vanguards of the new "natural" and organic movements seem like iconoclasts.[105]

Status Anxiety

The one theory that explains the emergence of all four ideals around 1880 and their decline around 1930 and the concentration of the trends in the urban upper-middle classes is that it was prompted by the dramatic growth in inequality toward the end of the century and its decline after the shocks of the world wars and the Great Depression. The growth in income inequality over the last three decades has attracted a great deal of attention since 2011, thanks in part to the Occupy movement, which rallied around the gap between the wealthiest 1 percent and the remaining 99 percent. However, like the four central ideals of the food revolution, the dramatic rates of inequality today are nothing new. The last time America's income distribution was as polarized as it is today was the turn of the last century. In terms of class structure, the twentieth century in America is bookended by these two periods of great inequality. In between, national income distribution converged in what Claudia Goldin and Robert Margo have called the great compression.[106] There are competing theories about its causes, but the typical narrative goes something like this: the runaway incomes of the robber barons and other Gilded Age elites plummeted during the Great Depression and then stagnated. Meanwhile, during and after the economic recovery driven by World War II, a more progressive tax structure and policies such as the minimum wage and the G.I. Bill directed more of the nation's growing wealth to wage laborers, creating a large and relatively prosperous middle class.

Thomas Piketty's analysis of the relationship between national capital stocks and economic growth helps clarify the extent of these shifts. For most of America's early history, inequality was relatively low compared to Europe. Land was plentiful and relatively cheap and the national stock of capital in America was relatively small, so new immigrants were able to catch up to their predecessors much more rapidly than workers could scale the social ladders of the Old World. As Alexis de Tocqueville wrote in 1840, "The number of large fortunes [in the United States] is quite small and capital is still scarce."[107] However, the picture changed significantly over the course of the nineteenth century.

As a small number of Americans began accumulate capital, primarily in the form of industry and more valuable real estate, the social classes diverged

and mobility between them declined. In 1810, the national wealth of the
United States amounted to barely three years of national income, compared
to more than seven years in most European countries. By 1910, U.S. national
capital was close to five years of national income—still lower than in Europe,
but the gap had shrunk by half in one century.[108] Land was still cheap and
plentiful, but with the rising stock of other forms of capital, the average small
landowner could no longer easily close the gap with the urban industrial
capital owner. The United States was coming to resemble France of La Belle
Èpoque or Charles Dickens's England. As Piketty notes, the shift was reflected
in the literature of the era. The novels Henry James set in Boston and New
York between 1880 and 1910 depict a city-dwelling East Coast elite "in which
real estate and financial capital matter almost as much as in European novels."[109] Piketty's chart shows that a high rate of income inequality persisted
until shortly after 1940 and that this was followed by four decades when the
richest 10 percent took home a far smaller share (i.e., the great compression).
After 1980, the rate of inequality returned to one that rivals that of the Gilded
Age (see Fig. 4).

The shifts in mainstream food culture parallel these shifts in inequality.
During periods of greater income inequality, the middle class relies more on
symbolic distinctions of taste to distance themselves from the lower class.
When the middle class as a whole is doing well materially and the upper class
isn't claiming as much of the nation's wealth, the cultural politics of taste shift.

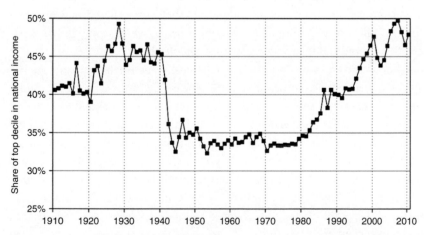

Figure 4. Income inequality in the United States, 1910–2010. Source: Thomas Piketty,
Capital in the Twenty-First Century, trans. Arthur Goldhammer, Copyright © 2014
by the President and Fellows of Harvard College. Reproduced by permission.

Then the middle class solidifies its identity by embracing the familiar and pedestrian and reacts more punitively against conspicuous displays of wealth or anything that smacks of class-climbing. If food trends are driven more by class anxieties than by increasing trade and knowledge, symbolic abundance, or proxies for feminism, that would also help explain the simultaneous emergence of contradictory trends such as French-inspired dinner parties with a different wine for every course and bland diets based on whole grains that eschew meat and alcohol.

Other scholars have speculated about how the growing gap between the emerging elite and most of the rest of the country might have affected popular culture. Levenstein explicitly connects the rise of French food in the Gilded Age to the changes in the U.S. class structure. According to him, the cuisine developed by the French upper class of the Second Empire was especially well suited to serve as a distinguishing form of conspicuous consumption for the American elite after the Civil War. Most of the ingredients used in French cooking were familiar and readily available, but the "elaborate methods of preparation, foreign code-words, and complex dining rituals" served as a "refuge from those trying to scale the ramparts of their newly acquired status."[110] Although French food had been associated with wealth and high social status since the colonial era, it was only in the Gilded Age that it became an object of desire for the masses. This is precisely the kind of cultural shift that an increase in inequality and decline in mobility would explain. For most of the previous two centuries, people who sought to scale the ramparts of America's social hierarchy could do so in a more direct manner through the income they might generate through their labor or by owning a small amount of land. Only toward the end of the nineteenth century, when a super-elite began to emerge and even professional incomes weren't sufficient to compete with the growing returns on capital, did the upper middle class begin to rely more heavily on cultural forms of distinction.

Hofstadter also attributes the rise of reform sentiment among urban professionals to the "upheaval in the distribution of deference and power" caused by the rise of a national super-elite. He argues that social movements such as the campaign for pure food were led primarily by Mugwumps—the college-educated class of doctors, lawyers, and business owners who tended to be local political leaders and active members in societies for civic betterment. In the last decades of the nineteenth century, they were eclipsed by the new financial elite: "In their personal careers, as in their community activities, they found themselves checked, hampered, and overridden by the agents of new corporations, the corrupters of legislatures, the buyers of franchises, the allies of political bosses. . . . In a strictly economic sense these men were not growing poorer as a class, but their wealth and power were being dwarfed by comparison with the new eminences of wealth and power. They were less important and they knew it."[111] Men who were used to being big fish in small ponds suddenly found themselves in an ocean full of much bigger fish. According to Hofstadter, they responded to the loss of moral authority by becoming active in reform movements that targeted the corporations that had disenfranchised them.

Hofstadter has little to say about the role that women played in those reform movements, but the status upheavals he describes would have affected women in the urban, professional classes in similar ways. The wives of doctors and lawyers, accustomed to being the chief arbiters of taste and morality in their communities, also found themselves displaced by the rise of a national super-elite they had little chance of entering. Their activism for pure food could also be seen as a way of lashing back at the new social order that had undermined them. Perhaps they were more inclined to perceive that new social order as threatening, or perhaps they exaggerated (or invented whole cloth) the dangers and excesses of industrial capitalism that they railed against in order to reassert their moral authority. It's worth noting that even African American women's clubs were composed almost exclusively of women in the emerging black middle class, and both their activism and their social functions (such as Japanese teas) have been widely interpreted as attempts to solidify their class identity.[112]

Shifts in class structure also offer a better explanation for the spread of weight-loss dieting than the theories of abundance or scientific rationality. Although the meaning of thinness transformed considerably over the period 1870 to 1900, changing from a sign of frailty and ill health to a sign of vigor and a youthful, modern beauty, it maintained an association with the upper class throughout that transition. Food didn't become more abundant and bodies did not become bigger, and thinness was not justified in rational terms until after rationality itself came into vogue. Slimming practices preceded, rather than followed, the popularization of the calorie and theory of energy balance. So the growing appeal of thinness seems more likely to have been a product of its associations with wealth than a product of its relationship to health. Striving for a slim body was yet another way for the stagnating middle classes to try to distinguish themselves from the masses.

The fact that the four ideals aren't ideologically consistent also makes status anxiety a particularly good explanation for their simultaneous rise. In the new social order that emerged in the 1880s, members of the professional middle class grasped at anything that would enable them to distance themselves from the lower classes, establish their capacity for conspicuous consumption, and assert a moral superiority over the robber barons who usurped them. It also explains their simultaneous decline. Pure food activism didn't end with the passage of the 1906 act or with woman suffrage. Industrialization actually increased throughout the period when that activism waned. What did change around the same time the four ideals began to decline was the structure of wealth and inequality in America. The Great Depression, two world wars, and progressive social policies enacted under Franklin D. Roosevelt reduced the stock of private capital dramatically.[113]

With inequality much lower and the middle classes flourishing economically, the incentive to pursue symbolic distinction diminished. French food did not become any less labor intensive or delicious, the logic underlying calorie math didn't change, whole grain diets didn't become any less virtuous, and the variety of ethnic and international foods that Americans had access to only increased. Nonetheless, the popular appeal of all of those things declined. As class structure changed, tastes changed. In the 1930s and 1940s, a new culinary ethos that idealized simplicity, frugality, reliability, and familiarity became dominant. Instead of idealizing foods that were marked by their distinction from the mainstream, the dominant culinary ideals of the mid-century celebrated all that was mainstream. Distinctive foods—whether they were gourmet, diet, natural, or foreign—were increasingly viewed with suspicion and disdain. As the Gilded Age embrace of these ideals receded from memory, the new culinary landscape dominated by meat and potato suppers set the stage for their revival in the 1980s. The intervening decades were just long enough to make the contemporary food revolution seem like a break with the past instead of a return to it.

CHAPTER 3

No Culinary Enlightenment

WHY EVERYTHING YOU KNOW
ABOUT FOOD IS WRONG

It ain't what you don't know that gets you into trouble. It's what you know for sure that just ain't so.

—Josh Billings

The conflicts between the four central ideals of the food revolution and their mainstream popularity during the Gilded Age certainly challenge the widespread belief that their resurgence since the 1980s is the result of a mass culinary enlightenment, but they don't undermine it completely. It's possible that the greater interest in new, different, and supposedly better foods really was the result of a rising tide of enlightenment and progress in both eras but that there was some kind of mid-century cultural regression, perhaps caused by the shocks of the world wars and the Great Depression. Maybe the similarities between the food trends of the two eras are merely a coincidence. Maybe aspirational food trends in the Gilded Age were largely driven by upper-middle-class status anxiety but similar trends today are actually enlightenment-driven this time around. Or maybe the best explanation for the trends in one or both of those periods is actually some hybrid of the culinary enlightenment thesis and status anxiety. For example, maybe the anxieties caused by high rates of inequality and stagnating middle-class wealth inspires people to adopt practices that really are better, even if they're also sometimes contradictory. The reason I don't think any of those things are true is because the justifications for most of the beliefs about food that have become common sense since the 1980s fall apart under even modest scrutiny.

Many commonsense beliefs about food start to sound strange when you try to make them concrete enough to test. The way most people believe in the superiority of gourmet food or thinness is qualified, not absolute. Even if you think expensive wine tastes better than the cheap stuff, you might also think there are great cheap bottles to be found or that there are real differences between expensive and cheap wines but only some people can perceive them. You might believe that most people in the overweight BMI range would be a little healthier if they lost a few pounds but also know some people who remain fat even though they seem to eat healthy, moderate diets and other people who are thin even though they eat a lot of junk food. I've tried to keep these qualifications in mind and not set up exaggerated, straw-man versions of popular food beliefs. This chapter summarizes the evidence that convinced me that much of what I thought I knew about food was either false or at least highly questionable. If the beliefs at the heart of this sea change in American eating habits are merely unsubstantiated, then it can't be true that people embraced them because of progress and enlightenment.

The Myth of the Discerning Palate

The ideas that taste perception is based primarily on the objective physical properties of food and that some people have a superior capacity to judge those properties are pervasive in contemporary food discourse. Books and classes on food promise to teach people how to improve their palates with the promise that it will help them enjoy food more and become more sophisticated consumers.[1] Some of the assumptions at work in the popular belief that palates can be trained are revealed by an episode of the show *Top Chef Masters* that featured a rare attempt to put the competitors' taste buds to the test.

Top Chef Masters is a spin-off of Bravo's hit cooking-competition reality show *Top Chef*. Unlike the original, which features mostly unknown but up-and-coming chefs, *Masters* features contestants who have already achieved a level of celebrity in their cooking careers, usually as the head chefs of successful fine dining restaurants in large cities. In season 1, episode 9, the final four chefs were asked to compete in a blindfolded taste test of twenty ingredients. All of them expressed apprehension about the task. Hubert Keller of Fleur de Lys, one of only seven restaurants in the Bay Area to receive a four-star rating from the *San Francisco Chronicle*, says, "The sense of sight for a chef is very important because you eat first with your eyes. It is a very difficult challenge." Michael Chiarello, chef and owner of Bottega in Napa Valley, claimed that he might be at a particular disadvantage: "I work with a very narrow scope of ingredients. I've been doing Italian, Italian, Italian. You slip me a little bowl of dashi and I couldn't even spell it, let alone tell you what it was."[2] These misgivings hint at problems with the ideal of objective taste perception that is implicit in a blind taste test: what we call taste is often based as much on visual cues and personal experience as it is on what goes on in our mouths.

One by one, the chefs were called into the kitchen. The host helped them put on a blindfold and then handed them small bowls filled with the test ingredients. Only a sampling of each chef's trials was shown. Three of the four correctly identified peanut butter, but Anita Lo, the first woman challenger to win on *Iron Chef America*, mistook it for tahini. Three of the chefs correctly identified ketchup and at least two of them got corn right. Chervil bedeviled Keller, who identified it as parsley, and Chiarello, who shook his head after tasting it and said, "I have no idea." Rick Bayless of President Obama's beloved Topolobampo mistakenly identified hoisin sauce as ranch salad dressing and mango as plum, after predicting, "I know when my chefs at home see this, they are going to make lots of fun of me." Finally, the host gathered the chefs to announce the results. Keller was in last place, having correctly identified five ingredients. "Not very many," he says, shaking his head sadly. The others did little better: Bayless and Lo tied for second with six correct guesses. Chiarello won with seven.[3]

The way the episode was framed and the way viewers reacted both demonstrate how the myth of the objectively discerning palate gets created and sustained, despite the fact that even experts like these master chefs aren't very good at identifying foods blind. The challenge itself is based on the assumption that being able to identify foods by taste alone is a meritorious skill and one that has some bearing on what makes someone a successful chef. There were real stakes: the winner of the challenge got first pick of sous chefs for the main event and their ranking factored into the elimination. The tasting trials the director chose to air in the episode was heavily biased toward correct guesses, which probably created the impression that the chefs were better at the task than they are. Altogether, the episode showed 28 of the 80 trials: 16 correct guesses and 12 wrong ones. Since the chefs had a total of 24 correct guesses and 56 wrong ones, a representative sample of 28 would have shown only 8 correct guesses and 20 wrong ones. Tellingly, although Keller was initially dismayed that he only guessed five ingredients correctly, once it turned out the other chefs hadn't done much better, their interpretation of the scores shifted. Upon hearing that she and Bayless both got six right, Lo said, "It was really nice to tie with Rick, who I know has a great palate, and is really an incredible chef, so I'm kind of patting myself on the back at this point." Just one guess better than Keller's "not very many" correct suddenly became evidence of having a "great palate." After Chiarello was announced as the winner, the other chefs congratulated him with newfound respect.[4]

Audience recaps of the episode also ignore the actual results of the test in favor of the narrative that successful chefs must have good palates and blind taste tests are a way to prove this. *Zap2It's* Sarah Jersild marveled at their poor performance, saying "I'm sure I'd suck at this, but I'd still hope that professional chefs would be able to ID hoisin sauce and peanut butter without too much trauma." She further suggested that Anita Lo's failure to identify those ingredients "makes me fear for her continued tenure on the show," implying that success in the blind taste test was a good metric of culinary skill.[5] *LA Times* blogger Krista Simmons wondered whether Keller's poor performance was a sign that "his sense of taste is dwindling with his old age,"[6] reasoning that he must have had a better palate at some point in order to get to where he is today. Multiple other bloggers were also "shocked" at Keller's loss, likely reflecting assumptions about the superior culinary skill and training required to cook French food.

The myth of the discerning palate helps sustain more rigid taste hierarchies and punitive social judgments about food that currently exist in other aesthetic realms, such as music or art. Having particularly good vision or hearing is rarely invoked as a factor in people's preferences in those realms, where criticizing other people's tastes has become less acceptable. As literature scholar Ashley Barnes writes in the *L.A. Review of Books*, "Today, as we all know, taste is a polyamorous affair, guilty pleasures have been declared innocent, and we are free to love whatever stirs us." Yet some of the same professors who might defend the value and cultural significance of reality television, Katy Perry, and *Fifty Shades of Grey* still decry junk food on aesthetic grounds as much as on nutritional ones—it's cloyingly sweet, too simple, or tastes synthetic. The idea that some people have good taste in food because they have better-trained taste buds provides a purportedly objective basis for the continued use of food as an arena of status competition.

Is there such a thing as an objectively delicious or disgusting food or is taste purely a matter of personal preference? The question speaks to a longstanding debate about the nature of aesthetics that begins with a basic conundrum. On the one hand, most people believe that taste is at least somewhat subjective. One person's favorite food might be repellent to someone else, even if they share the same background. On the other hand, most people also believe that some people's judgments are better than others; this belief is captured in the notion of good taste. While most people believe they personally have good taste, there's no necessary correspondence between someone's individual predilections and their sense of the broader social standards. For example, you might personally find bananas disgusting and yet not think that anyone who happens to like them has bad taste. Or you might really enjoy a scorned food such as American cheese, knowing your affinity is at odds with what counts as good taste in the dominant culture today. So if it's not just up to each individual to decide—if beauty isn't just in the eye of the beholder—where does the idea of good taste come from? Most attempts to defend the idea of objective taste standards appeal to either authority or democracy.

David Hume's 1757 essay "Of the Standard of Taste" is a classic example of the appeal to authority. He argues that all humans are similar enough that we're predisposed to enjoy the same things, but we end up with a variety of preferences primarily because of our differing capacities of discernment, or what he calls delicacy. By way of illustration, he recounts a story from *Don Quixote* in which two men who are supposed to be wine experts are asked to pass judgment on a cask of wine. One says it's good except for a faint taste of leather, the other says it's good except for a faint taste of iron. The rest of the company ridicules them, saying it just tastes good. But the experts are vindicated when they finish the wine and discover a key on a leather strap at the very bottom of the cask.[7] The rubes who liked the wine didn't really have different taste preferences than the experts—it's not that they like the taste of iron or leather—they just lacked the delicacy to detect the defects. According to Hume, delicate people are "sensible of every beauty and every blemish," and he says that they are the only people qualified to pass judgment on whether any work of art lives up to "the true standard of taste and beauty."[8]

Hume appeals to democracy when he explains how to distinguish between qualified critics and "pretenders." The appeal to authority is based on the premise that particular qualities—whether they be traces of iron or leather in wine or clearness of poetic expression—are universally offensive or pleasing. Not everyone can discern those qualities (although Hume suggests that most people can get better with practice), but supposedly they can collectively agree on who the most qualified critics are: "Some men in general . . . will be acknowledged by universal sentiment to have a preference above others." In another nod to consensus, Hume says that if you don't have concrete proof like the key in the cask, you can tell who the real experts are based on how well their preferences match up with majority opinion: "Wherever you can ascertain a delicacy of taste, it is sure to meet with approbation; and the best way of ascertaining it is to appeal to those models and principles which have been established by the uniform consent and experience of nations and ages."[9] If everyone ultimately likes and hates the same things, people with heightened perceptions are just better at picking out what most people will naturally like better, so the authoritative standard of taste will effectively be a democratic one, too. In another classic example of this line of reasoning, Immanuel Kant argues in *The Critique of Judgment* that there's a common sense (*sensus communis*) of beauty and everyone would converge on the same opinions about it if they all had enough education and leisure time to cultivate good reason, imagination, and perception.[10]

However, the story about the key in the wine also illustrates the inherent tension between a standard of taste based primarily on expert opinion and one based primarily on consensus. The majority of the company drinking the wine cannot taste the iron and leather and they all say it tastes good. For a truly democratic standard of good taste, the company would be correct and the experts' objections either irrelevant, or worse, a kind of deviance from the plurality's preferences. What Hume calls delicacy prevents the experts from experiencing the wine as the majority does, which could be seen as perverting rather than enhancing their judgment. Is delicacy really grounds for approbation? Are delicate people better at judging what everyone will like? And if not, if taste experts and the majority disagree, whose judgment ought to be the standard for good taste?

Today, the wine experts from *Don Quixote* might be called supertasters, a term psychologist Linda Bartoshuk coined in the 1990s. According to Bartoshuk, these people make up approximately 25–30 percent of the population. They detect more tastes and taste them more intensely than average tasters (who make up 40–50 percent of the population) or nontasters (the remaining 25–30 percent).[11] Thanks largely to a greater density of taste buds, supertasters are especially sensitive to the bitter compounds in green vegetables, coffee, and alcohol. Some avoid those things entirely or stick to a limited set of foods, often dominated by meat and starches. If you know any adults who seem to survive almost entirely on chicken fingers and French fries and swear they can tell if a piece of lettuce so much as grazed their plate, they may well have more taste buds than the average person. The range of taster types challenges both Hume's assumption that particular qualities in food will be universally offensive or pleasing and his assertion that discerning people make the best critics. A salad that might taste pleasantly bitter to an average taster and not bitter at all to a nontaster may well taste like poison to a supertaster. And as their picky tendencies suggest, while supertasters might have an advantage when it comes to detecting subtle tastes other people would miss, they're not necessarily very good at identifying which foods a majority of people would find pleasing.

In addition to where someone falls on the continuum of taster types, taste perceptions are affected by other genetic variations and life experiences. Flavor perception, which is dominated by the olfactory senses, can be altered by colds and allergies that affect the sinuses. Pregnancy sometimes enhances the sense of smell, which contributes to the development of new food aversions and affinities. Ear infections, head injuries, oral surgeries, and diseases such as Parkinson's and Alzheimer's can reshape taste by altering the taste buds and other sensory organs, damaging the nerves that carry information to the brain, or changing the brain itself.[12] Even if their tastes haven't been affected by illness or injury, most people have at least a few sensitivities, such as the genetic predisposition (unrelated to supertaster status) that makes approximately 15 percent of the population think cilantro tastes like soap. People can also develop conditioned aversions to foods consumed prior to a bout of vomiting. One bad night with tequila can make the smell repellent for life.

Taste perception can also be improved by repeated exposure, perhaps confirming Hume's belief that experience can enhance someone's inherent delicacy. However, exposure doesn't just improve your palate; it changes it. Many of the bitter, spicy, or pungent foods that are seen as acquired tastes are almost universally off-putting until people develop conditioned associations between them and the rewards they deliver. The process of taste acquisition depends on people's taster type and sensitivities as well as on cultural and personality factors. A lot of supertasters will never be able to distinguish between an Islay and Highland scotch because their aversion to alcohol makes them less likely to ever acclimate to its bitterness and heat. For people who can tolerate the taste of scotch, some will have easier access to the kinds of social settings where scotch is consumed and the potential rewards of learning to appreciate and talk about different regional styles, distilleries, and ages. The more scotch someone drinks, the more they're likely to be able to detect nuances of flavor and develop personal preferences, both of which will continue to reshape their taste perceptions.

To complicate matters further, taste perceptions are powerfully influenced by contextual cues such as color, plating, and price. An experiment from the 1970s that Eric Schlosser described in *Fast Food Nation* demonstrates the powerful effects of color on how people perceive food: "People were served an oddly tinted meal of steak and French fries that appeared normal beneath colored lights. Everyone thought the meal tasted fine until the lighting was changed. Once it became apparent that the steak was actually blue and the fries were green, some people became ill."[13] Less dramatically, in a 2004 study at the Cornell University Food and Brand Lab, thirty-two participants were invited to taste what they were told was strawberry yogurt. The lights were turned out and they were actually given chocolate yogurt. Nineteen of them still rated it as having "good strawberry flavor."[14] Color may also influence the taste differences many people claim to perceive in free-range or pasture-raised eggs. According to poultry scientist Pat Curtis, the egg industry has been conducting tasting experiments for years using special lights to selectively mask the color of the yolks. They reliably find that "if people can see the difference in the eggs, they also find flavor differences, but if they have no visual cues, they don't."[15]

Multiple studies have shown that presentation affects people's evaluations of food, too. Researchers at Montclair State University asked students to taste plates of hummus and chicken salad. The students given "neat" plates, with dollop of hummus centered on a lettuce leaf surrounded by carefully placed pita chips, carrots, and tomatoes or a perfectly round scoop of chicken salad, said they liked the food a lot more (rating it 30–37 points higher on a 200-point scale) than those given "messy" plates with the pita chips and tomatoes randomly scattered around a smear of hummus or an uneven pile of chicken salad.[16] Another study suggests that the critical quality may not be neatness but perceived artistry. In a study led by chef Charles Michel, who studied under Paul Bocuse in France and now does research with the Crossmodal Research Laboratory at Oxford University, sixty participants were served a salad composed of seventeen components presented in one of three styles: an "artistic" presentation inspired by Wassily Kandinsky's *Painting #201*, a "regular" presentation in which the components were mixed together and placed in the center of the plate, and a "neat" presentation with the compo nents placed side by side in rows without touching. People who received the "artistic" presentation rated the salad as significantly tastier than those who received either the "regular" or "neat" plates (7.5 on a 10-point scale, compared to 5.8 and 5.4, $p < .01$) and said they'd be willing to pay about twice as much for it (£4.25, compared to £2.08 and £2.14, $p < .01$).[17]

Many studies have also found that people report enjoying food and drink more if they believe it costs more. Brian Wansink, the same researcher who tricked people into thinking that chocolate yogurt was strawberry-flavored, found that people who paid $8 for an all you can eat buffet rated the food 11 percent higher on taste than those who paid $4.[18] Participants who tasted the same kind of chocolate said they liked it more when they were told it cost $15 than when they were told it cost $1.50. They also say they like it more when they're told it comes from Switzerland than when they're told it comes from China.[19] People typically expect wine that's more expensive to taste better,[20] even though in blind tastings they can't differentiate it from the cheap stuff. At the 2011 Edinburgh International Science Festival, psychologist Richard Wiseman asked 578 people to taste unlabeled samples of wine that ranged in price from £3.49 to £29.99 and guess whether it was under £5 or over £10. People guessed correctly 53 percent of the time for white wines and 47 percent of the time for red wines. It was a coin flip.[21] In fact, when they don't know the price, people may have a slight preference for cheaper wines. A massive review study concluded that "in a sample of more than 6,000 blind tastings, we find that the correlation between price and overall rating is small and negative, suggesting that individuals on average enjoy more expensive wines slightly *less*."[22]

However, if people believe a wine is expensive, they will not only say they like it better, their neurological response to it will also be different. A team of researchers led by economist Antonio Rangel at Cal Tech had twenty subjects who were screened for liking and at least occasionally drinking red wine taste five samples of Cabernet Sauvignon and a neutral control solution while using an fMRI to scan their brain activity. The wines were identified to the subjects only by their retail price, and two of the wines were administered twice. Wine 1 was presented at both $90 (its actual retail price) and $10, wine 2 was presented at both $5 (its actual retail price) and $45, and wine 3 was identified only at its actual retail price of $35. Unsurprisingly, almost everyone reported liking wines 1 and 2 more at the higher prices. The more interesting finding was that the higher prices also prompted more neural activity in an area of the brain thought to be involved in the experience of pleasantness.[23] When people expect a wine to taste better, as they generally do when they believe it costs more, they may genuinely derive more enjoyment from it.

None of the studies discussed so far involved people with any particular claim to what Hume called delicacy, but it turns out that experts with specially trained palates are no less subject to contextual cues. Frédéric Brochet demonstrated this in a series of experiments he conducted as a doctoral student at the University of Bordeaux involving subjects recruited from the Faculty of Oenology—that is, current and future wine-tasting experts. In one experiment, Brochet asked fifty-four of these expert tasters to describe wines in two sessions. In the first, they were given a red wine and a white wine and were asked to draw up a list of odor descriptors for each, using words on a provided list or their own terms. The words used most often to describe the white were honey, citrus fruit, floral, passion fruit, butter, and pear; the red was described with words such as wooded, spice, black currant, strawberry, cherry, prune, raspberry, vanilla, pepper, animal, and licorice.[24] In the second session, the same subjects were given two glasses of the same white wine used in the first session, one of which had been dyed red with a flavorless, odorless grape extract.[25] Each subject was given the list he or she had generated in the first session and asked to indicate which of the two wines most intensely indicated each descriptor. Despite the fact that the wines were identical, the tasters consistently associated the "red" wine with the descriptors they had used for the real red wine a week earlier while still pairing the undyed glass with their white descriptors.

In another experiment, Brochet served fifty-seven sommeliers the same red wine in two different sessions, first identifying it as a lowly table wine and then as a high-prestige Grand Cru Classé (the highest designation for a Burgundy vineyard). When the wine was identified as a table wine, it was described with largely negative words such as unbalanced (83 percent), fault (70 percent), weak (75 percent), simple (100 percent), flat (68 percent), sting (79 percent), and none (100 percent). However, when it was identified as a Grand Cru, the tasters used mostly positive words such as balanced (65 percent), agreeable (79 percent), full (87 percent), complex (73 percent), round (100 percent), and excellent (100 percent). Forty of the experts said the wine was worth drinking when it was labeled as a Grand Cru; only twelve said it was worth drinking when it was labeled as a table wine.

Another set of studies on wine experts directly challenges Hume's claim that discerning people will converge on certain universal preferences. Robert Hodgson, a retired oceanographer turned winemaker, has been running an experiment at the California State Fair wine competition to evaluate the consistency of the judges' scores since 2005. Each judge tastes a flight of thirty wines and Hodgson makes sure they get at least one sample in triplicate, poured from the same bottle. On average, the scores of the three identical samples vary by plus or minus 4 points out of 100, which is enough to significantly change its ranking. For example, typical judges might give the wine a 90 (Silver) the first time they tried it, an 86 (Bronze) on their second tasting, and then a 94 (Gold) on the third pass. Only about 10 percent of the judges gave all three of the identical samples scores within a range of plus or minus 2 points. Does this mean that 10 percent of the judges have really great palates and the rest are just pretenders? Apparently not. Hodgson has found that the judges who were consistent one year tend to regress back toward the mean in the next, suggesting that their consistency happens purely by chance.[26] Hodgson, who has since developed a test to determine whether any judge's assessment of a blind-tasted glass of wine in a medal competition is better than random chance, told the *Guardian* in 2013 that his preliminary experiments using the test have not been promising: "So far I've yet to find someone who passes."[27]

Brochet's and Hodgson's experiments have been used to cast aspersion on professional wine criticism. If wine experts can't even reliably tell red from white or recognize when they're drinking the exact same wine if it's served in different bottles, how much stock should anyone put in their evaluations? However, some people argue that cues such as color and price are inherent parts of the taste experience and that it's not only inevitable but completely valid to be influenced by them. If people can't tell the difference between red and white wines or vibrant and pale egg yolks unless they can see those differences, perhaps that's an argument against blind tastings, not against the validity of unblinded ones. Perhaps color and price are among the qualities people should use to judge the quality of their food.

But that's probably not what Hume had in mind when he described the principles of taste as universal and called for judgment based on "delicate sentiment, improved by practice, perfected by comparison, and cleared of all prejudice."[28] Instead, what the research on taste perception suggests is there's really no such thing as objective taste. Cilantro doesn't objectively taste like soap or not taste like soap—it does to some people and not to others. Scotch doesn't objectively taste like poison or peat—it might taste like one or both or neither, depending on your particular biology and biography. If you expect an egg to taste richer or more eggy when it's yellower, a yellower egg probably will taste that way to you whether the color comes from more carotenids in the yolk, warmer lighting in the room, or the addition of yellow dye. When it comes to taste, everyone is prejudiced—by their tongues, their genes, their experiences, and everything about the context in which they experience food, from price tags to plating. And yet Hume's parable of the key in the wine reflects how many people continue to think about good taste and who has it.

The fact that there is no such thing as objective taste undermines the theory that the growing interest in gourmet food since the 1980s is the result of people coming to the collective realization that premium foods taste better. For one, it's not clear that they do taste better, either to taste experts or to the masses. Many foods that are generally considered to be gourmet, such as pungent aged cheeses, sharp mustard, raw fish, drier wines, and mouth-puckeringly hoppy beers are off-putting to many people. On the other hand, many widely beloved foods such as Kraft macaroni and cheese, Doritos, and McDonald's French fries are excluded from gourmet status. Gourmet food is defined not by what is "universally pleasing," which in any case turns out to be probably nothing, but instead by a particular construction of good taste based primarily on scarcity. Enlightenment cannot explain the growing interest of the middle class in cultivating tastes distinguished by their scarcity. A more plausible explanation is that increasing inequality and stagnating class mobility has increased the anxieties and rewards associated with symbolic status competitions, driving more people to cultivate the tastes that carry elite cultural capital.

THE UNCERTAIN AND ELUSIVE HEALTH BENEFITS OF THINNESS

Although the Gilded Age idealization of thinness was initially more about beauty than about health, the medical justifications that were subsequently developed for the superiority of thinness have become a critical foundation for the culture of weight-loss diets. Especially since the 1980s, the growing concern about obesity has been justified by a set of interrelated beliefs about the relationship between body size, health, and food choices. First, size is seen as a proxy for health: in general, the thinner the better. Second, body weight is seen as primarily, if not exclusively, determined by behaviors thought to be within most people's personal control: eating the right foods in the right amounts and getting regular exercise. Third, it follows that thinness is understood to be something virtually everyone can and should achieve. It has become a moral mandate to be what is often referred to as a "healthy weight." This set of beliefs has become almost unassailable common sense: everyone knows that fatness is unhealthy and that the way to be thin is to eat the right foods in moderation and to exercise.

Nonetheless, there is considerable debate about what foods are the right foods. Packaged snack cakes and extruded corn snacks, any food from fast food restaurants, and sugar-sweetened beverages are universally thought to be fattening. The energy balance paradigm demonizes high-calorie, high-fat foods and large portion sizes.[29] However, alternative paradigms represented by low-carb, paleo, and plant-based diets have prompted considerable debate about the nutritional status of foods such as butter, eggs, full fat dairy, meat, bananas, potatoes, and whole grains. Green vegetables are about the only thing nearly everyone agrees is a healthy choice. Despite their differences, virtually all contemporary diet paradigms are invested in the overarching set of beliefs about diet and the body that amount to a meritocracy of thinness. A person's body size is seen as something they earn and deserve, a transparent indicator of how much prudence and willpower they have when it comes to food.

Almost a century of research on dietary interventions aimed at producing long-term weight loss suggests that the meritocracy of thinness is not just an exaggeration or an oversimplification of some basically valid biological facts. It does more than merely exclude a few outliers or underestimate the influence of genetic or structural factors that affect body size; it contradicts the findings of the overwhelming majority of research on weight-loss dieting and the majority of weight-loss dieters' personal experiences.

Fatness Doesn't Kill

Most claims about the risks of obesity are based on associations between body mass index (BMI) and the rates of death or disease in large epidemiological data sets. The first problem with this kind of evidence is that BMI is a

Figure 5. Christine Ohuruogu at the 2012 Olympics, overweight according to the BMI. Source: Photo by Jamie McPhilimey, originally published in Rick Broadbent, "Christine Ohuruogu Provides an Olympic Spark on Home Turf," *The Times* [London], July 24, 2013.

poor measure of how fat people are (in the sense of how much adipose tissue they have) because it's simply the ratio between their weight and the square of their height. It offers no information about their relative proportions of bone, fat, and muscle. Additionally, as a two-dimensional measure, it doesn't scale accurately for three-dimensional bodies. Thus, athletes and taller people are more likely to have a high BMI even if they have relatively low body fat percentages.[30] Based on current guidelines from the World Health Organization (WHO) and the National Institutes of Health (NIH), a BMI lower than 18.5 is underweight, 18.5–24.9 is normal, 25–29.9 is overweight, 30+ is obese (which is further divided into grade 1 for 30–34.9, grade 2 for 35–39.9, and grade 3 for 40+). According to the heights and weights of players published by the NBA and the NFL in 2005 and 2004, respectively, nearly half of professional basketball players and almost all professional football players are at least overweight, and over half of the football players are obese.[31]

The phenomenon isn't limited to linebackers or men. Track star Christine Ohuruogu, who won a silver medal in the 400-meter dash at the 2012 Olympics, is right in the middle of the overweight range with a BMI of 27 despite having virtually no visible body fat (see Fig. 5). BMI isn't a very good measure of fatness in non-athletes, either. The National Health and Nutrition Examination Survey (NHANES) III[32] collected the height, weight, and body fat percentage of 15,864 Americans in the period 1988–1994 (a nationally representative cross-section that excluded children under twelve and pregnant women). Overall, body fat percentage explained only 40 percent of the variability in BMI. Among participants with a BMI of 25, percentage of body fat ranged from 13.8 percent to 42.8 percent.[33]

The second problem with associations between BMI and death and disease is that the relationship they portray between size and health is complex; risks don't simply increase in a linear fashion with increasing weight. While the very fat generally die slightly younger than average, so do the very thin, and the longest lifespans are associated with the BMI range currently defined as overweight. This is neither a new nor controversial finding. A comprehensive review of the available data on weight, height, and mortality collected from life insurance policyholders in 1980 concluded that the group with the smallest percentage of deaths or "minimum mortality" was 10 to 20 percent heavier than the weights recommended by MetLife's "ideal weight" tables and increased with age.[34] Since then, nearly every study and review of the literature on body size and mortality has supported the conclusion that overweight people live longer than those in the normal BMI range.

According to a study published in the *Journal of the American Medical Association* in 2005 that examined the BMI and mortality data of over 36,000 Americans collected in NHANES I, II, and III, people in the overweight category consistently had the longest average lifespans. Mild obesity (grade 1) was associated with a small increase in the risk of dying within the ten-year follow-up period, but the risk of death for underweight people was higher. For people over the age of sixty, the risk associated with being underweight was greater than the risk associated with extreme obesity (grades 2 and 3).[35] These findings aren't unique to the United States. In a study that tracked 1.8 million Norwegians for ten years starting in the 1970s, the highest average life expectancy (79.7 years) was found in the group with a BMI of 26 to 28 (overweight) and the lowest average life expectancy was found in the group with a BMI below 18 (underweight). People with a BMI of 18 to 20, most of whom would be categorized as normal, had a lower average life expectancy than those with a BMI of 34 to 36 (obese).[36] Studies of 170,000 people in China, 20,000 people in Germany, and 12,000 people in Finland have found the same thing: overweight people tend to live as long or longer than people in the normal weight range.[37]

Premature mortality risk does increase with increasing body weight after a point. When graphed, the relationship between mortality and body size looks like a U-shaped or J-shaped curve with the lowest risk (or longest lifespan) in the overweight range and higher risks for the underweight and the very obese. In most studies, the additional mortality associated with fatness does not become statistically significant until well beyond the BMI threshold for obesity. In 1996, a team led by nutritionist Richard Troiano conducted a meta-analysis of nineteen longitudinal studies of BMI and mortality that concluded that mortality was highest among the very thin and the very heavy. On the heavy side, increased mortality remained within the bounds of statistical uncertainty until a BMI of 40 or more. The researchers concluded, "This analysis of mortality suggests a need to re-examine body weight recommendations. Weight levels currently considered moderately overweight (i.e. a BMI > 27) were not associated with increased all-cause mortality" and suggested that "attention to the health risks of underweight is needed."[38] Ironically, the NIH cited the findings of Troiano and colleagues in a report on its 1997 decision to lower the BMI threshold for overweight from 27.8 for men and 27.3 for women to 25 for everyone, despite the fact that the study directly contradicted the report's recommendations.[39]

In an even larger meta-analysis published in 2013 that included ninety-seven studies with a combined 2.88 million individuals and over 270,000 deaths, being overweight was again associated with a lower risk of premature death. Grade 1 obesity was associated with no additional risk of premature death compared to the risk for those in the normal weight range. Only among people with a BMI over 35 was there any statistically significant relationship between fatness and premature death.[40] The data offer no indication that the vast majority of overweight and obese Americans who have a BMI lower than 35—approximately 130 million of the roughly 165 million Americans in those pathologized categories—have any additional risk of premature death.[41]

The third problem with BMI-based associations is that correlations are limited in terms of what they can tell us about the relationship between weight and mortality. While people classified as underweight and those in obese grades 2 and 3 appear to have a slightly increased risk of premature death, it's unclear if that's because of their weight. Even controlling for other factors, there is no way to completely rule out the possibility that the associations are spurious or that weight extremes are merely proxies for other causes of death. And many studies fail to control for one or more of the factors that are likely to be confounding variables in the relationship between weight and mortality, such as race, income, insurance status, smoking, activity levels, and known health conditions. One particularly egregious example is the 1999 Centers for Disease Control (CDC) study led by David Allison that gave rise to the oft-repeated claim that obesity causes 300,000 deaths per year in the United States. Allison and his colleagues used the data from several large epidemiological studies to come up with estimates of how many of the total deaths in the year 1991 were associated with each BMI category. Then they assumed that all additional estimated mortality in the "obese" group relative to the "normal" group was caused by the difference in weight. Not only did they fail to control for factors such as race and income, which are correlated with both higher weights and higher mortality risk, but they made no distinctions between deaths from causes potentially related to obesity (such as heart disease) and ones that are probably unrelated (such as car accidents).[42]

In 2005, another CDC team led by Katherine Flegal published a revised estimate based on the deaths in 2000. This time researchers accounted for confounding factors such as age, gender, smoking status, race, and alcohol consumption. They found that obesity was associated with an estimated 111,909 additional deaths compared to the normal BMI category (18 to 25) and that being merely overweight was associated with an estimated 86,094 fewer deaths. They also found that being underweight was associated with an extra 33,746 deaths, even though the people in this category constituted a much smaller percentage of the population than the obese. By the same logic that Allison's team used to attribute all additional deaths among the obese to weighing too much, one could conclude that nearly as many people in the normal category (and more if you include the underweight) are dying from weighing too little as the number that is dying from being obese.[43] Ultimately, this kind of data can't reveal the underlying cause of the association. Mortality rates could be higher among the very thin and the very fat because weight gain or loss is often among the symptoms caused by life-threatening diseases. The very fat might be more likely to die younger because they're less likely to seek medical care as a result of the documented stigma and prejudice they face from doctors and nurses.[44] Even studies that control for factors such as age and income cannot rule out the possibility that the association between weight and additional mortality or disease might be caused by a confounding variable.

Similar issues plague much of the evidence on the relationship between weight and disease. It's true that higher BMIs are associated with an increasing incidence of diabetes, heart disease, osteoarthritis, and some forms of cancer. However, once again, these associations are based on correlations, and many studies on the relationships between weight and disease fail to take factors such as smoking, income, race, insurance status, activity levels, and preexisting conditions into account. In most cases, the correlations are weak and there is no good explanation for why additional body mass or fat tissue would cause the associated condition. For example, according to the WHO, high blood pressure is two to three times more common among obese people than people in the normal BMI ranges.[45] That may sound large, but when a risk factor has a causal link to a disease, the association is typically much larger. For example, people who smoke are fifteen to thirty times more likely to get lung cancer than people who do not smoke.[46] Most of the epidemiologists interviewed by *Science* magazine in 1995 for a special report on risk factors said they would not take seriously an association with smaller than a three- or four-fold increase in the disease it was suspected of causing in a large, well-controlled study.[47] Even meeting that bar is not sufficient to prove a causal relationship, but it would at least merit further investigation.

Some evidence suggests that the association of obesity with high blood pressure (or hypertension) may be at least partially explained by the effects of weight cycling, or losing and then regaining weight. In both rat studies and human studies, subjects who lost weight on calorie-restriction diets and then regained most or all of the weight they lost had higher blood pressure than subjects who weighed the same as the rats or people who had weight cycled but had not lost and regained a significant amount of weight.[48] That finding might explain why the relationship between obesity and high blood pressure is weaker in cultures where dieting is less common, such as American Samoa and rural China.[49] It might also explain why hypertension increased in Europe following World War II as people whose diets had been restricted due to wartime food shortages regained the weight they had lost.[50] Another possibility is that high blood pressure in a thin person might actually be normal for an obese person, who has more blood and a larger body to pump it through.[51] Numerous studies have shown that obese people with high blood pressure live significantly longer than thinner people with high blood pressure, one of the "obesity paradoxes."[52] Of course, if obesity itself is not a cause of disease or death, there is no reason to consider better outcomes in fat people to be paradoxical.

Obese people with type 2 diabetes also live longer than thinner people with the condition.[53] Unlike most other diseases associated with fatness, type 2 diabetes is strongly associated with increasing BMIs. According to the data from the first fifteen years of the Nurses' Health Study, which follows over 100,000 registered nurses who were between the ages of thirty and fifty-five in 1976, women with a BMI of 22–22.9 (well below the threshold for being overweight) were approximately three times more likely to develop type 2 diabetes than women with a BMI lower than 22, women at the low end of the overweight range (BMI between 25.0 and 26.9) were eight times more likely to develop type 2 diabetes, and obese women were forty times more likely to develop type 2 diabetes.[54] These associations are large enough to suggest that there is probably a meaningful relationship between BMI and diabetes. However, there is ongoing debate about whether fatness is more likely to be the cause of diabetes or the result of it.

The theory that diabetes might cause weight gain rather than the other way around is sometimes called Bennett's model: genetic and environmental factors cause some people to develop elevated insulin levels, which causes insulin resistance and weight gain. The additional body fat may exacerbate the development of insulin resistance, but it is better understood as a symptom of the type 2 diabetes that eventually develops than the cause of the condition.[55] Subsequent research has shown that elevated insulin precedes weight gain in many people who go on to develop type 2 diabetes and that people with a family history of diabetes have higher insulin levels regardless of their weight.[56] Gaining weight, particularly fatty tissue, may actually be an adaptive or protective response to elevated insulin levels, which could explain the "obesity paradox" for diabetes. The potentially protective role of fat may also explain why some studies have found that type 2 diabetics who lose weight have up to a threefold increase in the risk of premature death compared to those remain overweight or obese.[57]

Dieting Doesn't Work

Even if it could be established that fatness is in fact the cause of premature death or the cause of the diseases it is correlated with, it wouldn't make sense to prescribe dieting as a therapy unless it was established that diet-induced weight loss would help rather than worsen the condition, as it seems like it might do for type 2 diabetes and hypertension. No one knows yet whether diet-induced weight loss has long-term health benefits for the simple reason that no diets have ever been shown to produce long-term weight loss in more than a small percentage of study participants. Virtually every study of weight-loss dieting that has followed participants for longer than six months has found that the vast majority of dieters ultimately regain all the weight they lose initially, if not more.

The first comprehensive review of the literature on weight-loss dieting was published in 1959 by Albert Stunkard and Mavis McLaren-Hume. It serves as a template for many of the review articles published since. First, they critiqued the poor quality of most diet experiments and the reporting on them. Of the "hundreds of papers on treatment for obesity" that had been published in the previous three decades, only eight met basic criteria such as specifying the duration of the study and reporting on the drop-out rate. Next, they summarized the consistent failure of studies to produce significant weight loss in most of the subjects, regardless of the type of treatment used: "The results of treatment for obesity are remarkably similar and remarkably poor. Thus, with the exception of Feinstein, no author has reported even the modest success

of a 20 lb. weight loss in more than 29 percent of his patients."[58] Across the
eight studies that reported enough data to enable a reasonable evaluation
of the results, only 5 percent of the subjects were able to lose forty pounds,
which may be the origin of the oft-cited statistic that 95 percent of weight-
loss diets fail.[59] Finally, Stunkard and McLaren-Hume attempted to account
for the persistent assumption that weight-reduction programs are effective,
in spite of the lack of evidence. They attributed it largely to the ambiguity in
how most studies report their results and the "naive optimism of the medical
profession."[60]

In Stunkard and McLaren-Hume's own experiment, they tracked the
results of all the patients referred to the Nutrition Clinic at New York Hospital
over a three-month period. This yielded a subject pool of 100 obese people
who were then defined as weighing at least 20 percent more than the ideal for
a person of their height and who were of "medium build" according to the
Metropolitan Life tables. Only three men were referred to the clinic during
that period, so the subject pool was made up almost entirely of women with
a median weight that was 44 percent higher than the ideal. For a five-foot,
five-inch woman, the desirable range was 116–130 pounds, so the study participants likely averaged approximately 177 pounds. The subjects were given
routine weight-loss advice that included instructions to follow "balanced
weight-reduction diets from 800 to 1500 Cal" and were asked to return to the
clinic every two to six weeks for the next two and a half years. The drop-out
rate was high; thirty-nine patients did not return for a second visit. Of the
subjects who continued treatment, only twelve lost twenty pounds or more at
any point. Of those, only six maintained a loss of at least twenty pounds for a
year, and only two maintained a twenty-pound loss for two years. As Stunkard
and McLaren-Hume note, this result replicates the findings of the few previous studies that measured long-term weight-loss maintenance, which consistently found that "a majority of persons regain a majority of the pounds lost."[61]

Fifteen subsequent review articles on weight-loss dieting have reached the
same conclusion.[62] In most studies, participants lose some weight initially,
but after about six months they begin to gain weight back. By twelve months
after the time a diet begins, most participants have regained a majority of the
weight they initially lost. In studies that follow participants for three years or
more, from one-third to two-thirds regain more weight than they initially
lost. In almost all studies, the number of people who gain back more weight
than they lost is higher than the number who maintain a majority of their
initial weight loss. Despite these dismal results and potential risks of weight
cycling,[63] researchers often mysteriously conclude that dieting promotes
weight loss and health.

In 1991, psychiatry professors David M. Garner and Susan C. Wooley set out to develop evidence-based recommendations for mental health practitioners seeking to treat overweight patients. After reviewing the literature, they decided to advise against prescribing dietary treatments for obesity on the grounds that none had proven effective. The evidence for the failure of dieting was so overwhelming that they shifted their focus to trying to explain why medical professionals continue to promote behavioral treatments for obesity despite decades of evidence showing them to be ineffective. They found that authors of the weight-loss studies tend to "advocate a critical attitude towards obese patients" and interpret the failure of behavioral treatment programs as "understandable only as a consequence of patient noncompliance." Ultimately, Garner and Wooley accused researchers of "blaming the victim" for their own ineffective treatments and called on health professionals to explore alternative approaches to addressing the "physical, psychological, and social hazards associated with obesity without requiring dieting or weight loss."[64]

In one of the studies that Garner and Wooley reviewed, researchers randomly assigned participants to three experimental groups: one received standard behavioral therapy (counseling about diet and exercise strategies aimed at creating a caloric deficit), one received a weight-loss drug, and one received both the therapy and the drug. All of the treatment groups lost a significant amount of weight in the first six months, and all of the treatment groups showed significant regaining by the end of the eighteen-month follow-up. But instead of concluding that all of the treatments had failed, the study's authors claimed that their results provide hope for behavioral therapy, because that group showed the slowest rate of regaining (see Fig. 6).[65] According to Garner and Wooley, this interpretation of the data is par for the course in dieting research. After four years, nearly all participants in nearly all studies gain back nearly all the weight they initially lost, yet the authors of those studies typically insist that the initial weight loss proves that the diet interventions are effective.[66] Sometimes the researchers claim that if the participants had not dieted, they would have weighed even more. In studies of standard treatments such as calorie restriction, exercise, and counseling, the authors often call for more aggressive treatments such as very low calorie diets (VLCD, or diets of less than 800 kcal/day) or supervised fasting, even after series of sudden, unexpected deaths in VLCD trials in the late 1970s raised concerns about their safety.[67] The belief that people can and should be thin if you can just put them on the right diet seems to be invulnerable to evidence.

A review published in 2000 by Robert W. Jeffrey and colleagues also described the pattern of weight loss and regaining among patients who participate in behavioral treatments for obesity as "remarkably consistent."[68]

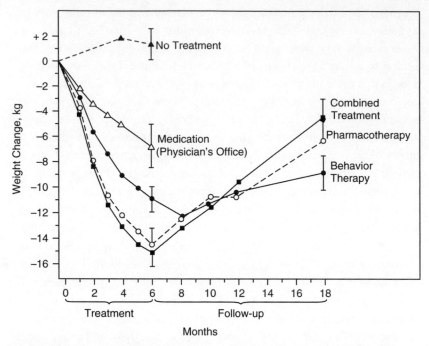

Weight changes during six-month treatment and12-month follow-up. The three major treatment groups lost large amounts of weight during treatment: behavior therapy (closed circles), 10.9 kg; pharmacotherapy (open circles), 14.5 kg; and combined treatment (squares), 15.3 kg. Behavior-therapy group continued to lose weight for two months and then slowly regained it; pharmacotherapy and combined-treatment groups rapidly regained weight. Among control groups, no-treatment (waiting-list) group (closed triangles) gained weight; physician's office medication group (open triangles) lost 6.0 kg. Patients in control groups received additional treatment at six months and so were not available for follow-up. Vertical lines represent 1 SEM.

Figure 6. Weight regain in a trial of behavior therapy and pharmacotherapy for obesity. Source: Linda W. Craighead, Albert J. Stunkard, and Richard M. O'Brien, "Behavior Therapy and Pharmacotherapy for Obesity," *Archives of General Psychiatry* 38, no. 7 (1981): 765.

A series of graphs from different studies repeatedly show a nadir between six months and one year of treatment followed by an increase back to the baseline (or higher in the studies with the longest follow-up) (see Fig. 7). The fact that the studies used a wide range of different strategies used to induce weight loss led Jeffrey and his colleagues to note that no matter what researchers do, most dieters achieve their maximum weight loss at six months and gradually regain all or almost all of the initial loss within three to five years.

Instead of concluding that long-term weight loss is biologically impossible for some people or that dieting is a poor strategy for achieving it, Jeffrey and

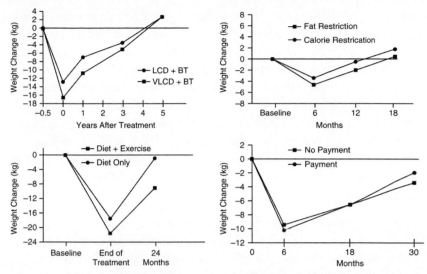

Figure 7. Long-term weight loss trends in (clockwise from top left) National Task Force on the Prevention and Treatment of Obesity 1993; Jeffrey et al. 1995; Sikand et al. 1988; and Jeffrey et al. 1993. Sources: Robert W. Jeffrey, Leonard H. Epstein, Terence G. Wilson, Adam Drewnowski, Albert J. Stunkard, and Rena R. Wing, "Long-Term Maintenance of Weight Loss: Current Status," *Health Psychology* 19, no. 1 (Supplement) (2000): 8–10; National Task Force on the Prevention and Treatment of Obesity, "Very Low-Calorie Diets," *JAMA* 270, no. 8 (1993): 967–974; R. W. Jeffrey, W. L. Hellerstedt, S. A. French, and J. E. Baxter, "A Randomized Trial of Counseling for Fat Restriction Versus Calorie Restriction in the Treatment of Obesity," *International Journal of Obesity* 19, no. 2 (1995): 132–137; G. Sikand, A. Kondo, J. P. Foreyt, P. H. Jones, and A. M. Gotto, "Two-Year Follow-Up of Patients Treated with a Very-Low-Calorie Diet and Exercise Training," *Journal of the American Dietetic Association* 88, no. 4 (1988): 487–488; R. W. Jeffrey, R. R. Wing, C. Thorson, L. R. Burton, C. Raether, J. Harvey, and M. Mullen, "Strengthening Behavioral Interventions for Weight Loss: A Randomized Trial of Food Provision and Monetary Incentives," *Journal of Consulting and Clinical Psychology* 61, no. 6 (1993): 1038–1045. Note: LCD+BT=low-calorie diet and behavior therapy, VLCD+BT= very-low-calorie diet and behavior therapy.

colleagues implicitly blamed the overweight subjects who supposedly failed to maintain the behaviors that caused their initial weight loss: "The experience of people trying to control their weight is a continuing source of fascination and frustration for behavioral researchers. Overweight people readily initiate weight control efforts and, with professional assistance, are quite able to persist, and lose weight, for several months. They also experience positive outcomes in medical, psychological, and social domains. Nevertheless, they almost always fail to maintain the behavior changes that brought them these positive results."[69] They imply that the dieters should be able to maintain the behaviors, and thus presumably the weight loss, without any evidence to support that assumption. Indeed, several studies have found that subjects begin to regain weight even while reportedly adhering to restrictive diets.[70]

Jeffrey's own experimental research attests to the remarkably consistent failure of weight-loss dieting. In 1995, he and Rena Wing published the results of one of the few large, long-term, controlled weight-loss studies ever conducted. They recruited 202 participants between the ages of twenty-five and forty-five who were between forty-one and seventy-one pounds above the MetLife ideal weight standards and assigned them randomly to one of five experimental groups:

1. A control group, which received no intervention.
2. A standard behavior therapy (SBT) group that participated in group counseling sessions once per week for the first twenty weeks and once per month thereafter, with weekly weigh-ins between sessions. Behavioral counseling included instruction on diet, exercise, and behavior modification techniques. Dietary goals were assigned at 1,000 or 1,500 kcal per day depending on initial body weight. Exercise recommendations were to walk or bike five days per week, beginning with a weekly goal of 250 kcal per week and gradually increasing to 1,000 kcal per week. Participants were asked to keep eating and exercise diaries regularly throughout the program.
3. Participants in the third treatment group, SBT + food, were given SBT and also were provided with food each week for eighteen months. Food consisted of premeasured and packaged dinners and breakfasts for five days per week.
4. The fourth treatment condition, SBT + incentives, consisted of SBT plus an incentive program through which each participant could earn financial rewards up to $25 per week for achieving and maintaining weight loss.
5. The last treatment group, SBT + food + incentives, included all of the treatment elements described earlier in combination (i.e., SBT, food provision, and incentives).[71]

The graph of their results, measured over the ensuing two and a half years, echoes the graphs Jeffrey and colleagues collected in their review article: a nadir in all the treatment groups at six months and steady regain afterward (see Fig. 8).

In addition to examining the subjects periodically throughout the eighteen months of the treatment, researchers contacted participants a full year after the study ended for an additional follow-up, which was completed by

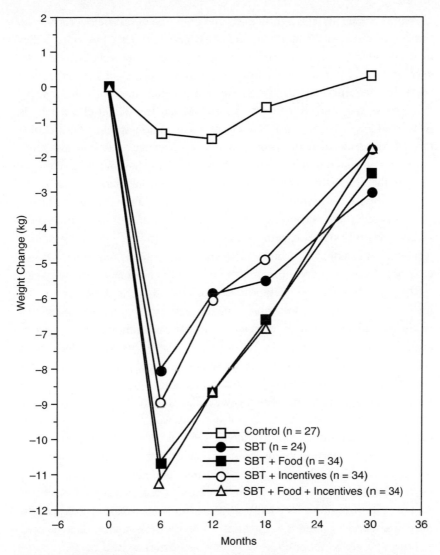

Figure 8. Long-term weight loss trends for standard behavioral treatment, food provision, and monetary incentives. Source: Robert W. Jeffrey and Rena R. Wing, "Long-Term Effects of Interventions for Weight Loss Using Food Provision and Monetary Incentives," *Journal of Consulting and Clinical Psychology* 63, no. 5 (1995): 795. Note: SBT = standard behavioral treatment.

177 (88 percent) of the original participants. By that point, there was no significant difference between any of the treatment groups and the control group. Jeffrey and Wing concluded that "the overall results of this evaluation reemphasize the important point that maintaining weight loss in obese patients is a difficult and persistent problem."[72]

The vast majority of weight-loss studies prescribe moderate calorie restriction, as in Jeffrey and Wing's "standard behavioral treatment," but some studies have also sought to compare different diet strategies. Low-carbohydrate diets, such as the Atkins diet, seem to lead to slightly greater weight loss in the short term, but the long-term results are not much better. In a 2003 study, sixty-three obese men and women with an average starting weight of 217 pounds were randomly assigned to follow either a low-calorie and low-fat diet or a low-carbohydrate diet; the people on the low-carbohydrate diet lost more than twice as much weight as those who followed the low-calorie and low-fat diets in the first six months (about fifteen pounds compared to about seven pounds). By twelve months, both groups had regained weight and the difference between them was no longer statistically significant.[73] In a 2007 study, 311 overweight or obese women with an average starting weight of 187 pounds were randomly assigned to follow either the Atkins (low carbohydrate), Ornish (low fat), Zone (30 percent protein; 30 percent fat; 40 percent carbohydrates), or LEARN (low-fat and lifestyle behavior change) diets. Once again, the low-carbohydrate group lost more weight in the first six months, and in this study, they also maintained more weight loss at twelve months (10.4 pounds compared to 5.7 for those who followed the Ornish diet, 3.5 for those who followed the Zone diet, and 4.9 for those who followed the LEARN diet) but all of the groups were regaining weight at that point despite the fact that they were still eating significantly fewer calories than before they started the diets. The researchers admitted that "longer follow-up would likely have resulted in progressively diminished group differences."[74] In a 2008 study, 322 participants with an average starting weight of 201.5 pounds were randomly assigned to follow a low-carbohydrate diet, a low-fat diet, or a Mediterranean diet for two years. All groups lost weight in the first six months and then started regaining. At twenty-four months, they had maintained an average loss of 10.4 pounds, 6.3 pounds, and 9.7 pounds, respectively.[75] Clearly, none of these diets succeed in bringing overweight or obese subjects into the normal BMI range.

Another review of the literature on weight-loss dieting was published in 2007 by a team of UCLA researchers tasked with developing recommendations for Medicare regarding obesity prevention. Like reviewers dating back to Stunkard and McLaren-Hume, the UCLA team lamented the quality of much of the published research on weight-loss dieting. They were able to find only seven studies of weight-loss dieting where participants were randomly assigned to diet or control groups and were followed for at least two years, which they define as the "gold standard" that is necessary in order to make causal claims about the long-term effects of dieting. Only three of the gold standard studies found statistically significant differences between the diet and control groups by the end of the follow-up, and even in those, the amount of weight loss maintained was too small to be medically significant. Across all seven studies, the average weight loss maintained was 2.4 pounds, ranging from a high of 10.4 pounds in the study with the shortest follow-up time to a gain of 3.9 pounds. The authors noted that "it is hard to call these obesity treatments effective when participants maintain such a small weight loss. Clearly these participants remain obese."[76]

Because of the relative paucity of controlled studies, the UCLA team also examined fourteen studies with long-term follow-up that didn't include control groups. The average initial weight loss in those studies was 30.8 pounds, but in the follow-up periods, participants typically gained back all but 6.6 pounds. Furthermore, an average of 41 percent of participants weighed more at the follow-up than before they went on the diet. Mann and colleagues described several problems with these studies, such as low participation rates in the long-term follow-ups, heavy reliance on self-reporting, and failure to control for the likelihood that some participants were already dieting again at the follow-up. However, all of those factors should have biased the results in the direction of showing greater weight loss and better long-term maintenance, so if anything the real performance of the diets is likely to be even worse.

Finally, the UCLA team looked at ten long-term studies that did not randomly assign participants to experimental and control groups. Most of these were observational studies that assessed dieting behavior and weight at the start of the study and then followed up with participants to ask about their self-selected dieting behaviors and measure changes in weight over time. Of those studies, only one found that dieting led to weight loss over time, two showed no relationship between dieting at the baseline and long-term weight gain, and seven showed that dieting at the baseline was associated with weight gain. As did Garner and Wooley in 1991, the UCLA team concluded that Medicare should not recommend weight-loss dieting as a treatment for obesity: "It appears that dieters who manage to sustain a weight loss are the rare exception, rather than the rule. Dieters who gain back more weight than they lost may very well be the norm, rather than an unlucky minority. If Medicare is to fund an obesity treatment, it must lead to sustained improvements in weight and health for the majority of individuals. It seems clear to us that dieting does not."[77]

There are, of course, exceptions. The dismal results of the weight-loss experiment Rena Wing collaborated on with Robert Jeffrey prompted her to start the National Weight Control Registry (NWCR) to collect success stories and data from people who have lost a minimum of thirty pounds and kept it off for at least one year. According to the NWCR website, they have over 10,000 members who have lost an average of sixty-six pounds and have kept it off for an average of 5.5 years.[78] The goal of the registry is to identify strategies that might help other dieters, but as the researchers who run the registry admit, "Because this is not a random sample of those who attempt weight loss, the results have limited generalizability to the entire population of overweight and obese individuals."[79] Surveys of the registry members show most of them follow a low-calorie and/or low-fat diet, although about a third follow a low-carb diet. A majority eat breakfast every day, weigh themselves at least once a week, watch less than ten hours of TV per week, and engage in regular physical activity.[80] In other words, they engage in exactly the kinds of behaviors that the participants in most weight-loss studies are counseled (or, in clinical settings, forced) to do. The NWCR has yet to figure out what makes registry members different from the vast majority of dieters.

One difference the more general statistics may fail to capture is an extraordinary commitment on the part of long-term weight-loss maintainers to chronically restrained eating and daily exercise. Describing NWCR members, Kelly Brownell, director of the Rudd Center for Food Policy and Obesity at Yale University, says: "You find these people are incredibly vigilant about maintaining their weight. Years later they are paying attention to every calorie, spending an hour a day on exercise. They never don't think about their weight." Several registry members described their weight-loss maintenance strategies in a *New York Times* article. Lynn Haraldson, a 48-year-old woman who weighed 300 pounds in 2000 and then dropped to 125 pounds on Weight Watchers, manages to stay around 140 pounds "by devoting her life to weight maintenance. She became a vegetarian, writes down what she eats every day, exercises at least five days a week and blogs about the challenges of weight maintenance. . . . She has also come to accept that she can never stop being 'hypervigilant' about what she eats."[81] Janice Bridge, a 66-year-old woman who weighed 330 pounds in 2004 and dropped down to 165 pounds on a medically supervised 800-calorie diet in 2006, says, "It's one of the hardest things there is. It's something that has to be focused on every minute. I'm not always thinking about food, but I am always aware of food. . . . My body would put on weight almost instantaneously if I ever let up." Bridge limits herself to 1,800 calories, drinks at least 100 ounces of water, and exercises from 100 to 120 minutes every day. She also weighs everything she eats and avoids all sugar and white flour. Despite these continual efforts, Bridge's weight has stabilized around 195 pounds.

Research on the entire database of NWCR members suggests that Bridge and Haraldson are not alone in their extreme vigilance. Ninety percent of registry members exercise for at least an hour every day. The average registry member logs 420 minutes of activity per week, over twice as much as the CDC's 2008 physical activity guidelines recommend for most adults.[82] Nearly 40 percent of NWCR members weigh themselves at least once a day.[83] Most continue to eat restricted diets even years after their initial weight loss. The average dietary intake reported by registry members is 1,685 kcal per day for men and 1,306 kcal per day for women, significantly lower than the averages for age-matched NHANES III participants (2,545 kcal per day for men and 1,764 kcal per day for women).[84] Many registry members report eating the same foods in the same patterns every day and refusing to "cheat" even on weekends or holidays.[85] Registry members score almost twice as high on a measure of restraint in eating as both normal-weight and obese nondieters (15.1 on a 21-point scale, compared to 8.1 for normal-weight nondieters and 7.0 for obese nondieters) and even higher than people diagnosed with bulimia nervosa (12.1).[86] Furthermore, even though registry members are unusually successful for weight-loss dieters, many remain overweight or obese. The average BMI of nearly 4,000 registry participants who enrolled from 1993 to 2004 was 25.[87]

The fact that maintaining significant weight loss, albeit not a normal BMI, appears to be possible for a small minority of dieters does not mean that it is possible for everyone. Somewhere between 45 million and 90 million Americans diet to lose weight every year, most of them by attempting to reduce their caloric intake.[88] According to a survey conducted in April 2010 by a private consumer research firm on behalf of Nutrisystem, 30 percent of Americans have dieted repeatedly, and the average number of attempts is twenty among repeat dieters.[89] According to a 2011 Gallup poll, nearly three-quarters of women and half of men have "seriously tried to lose weight" at some point in their lives; women make an average of 7 attempts and men try an average of 3.6 times.[90] If dieting made fat people thin, there would be far fewer overweight and obese Americans.

The failure of most weight-loss diets makes it essentially impossible to evaluate whether sustained weight loss would improve most fat people's health. However, several observational studies suggest that weight loss *increases* the risk of premature death among obese individuals, even when the weight loss is intentional. According to an analysis of the NHANES III data set, mortality for people over the age of fifty was higher for people who lost 5–15 percent of their starting weight than those in the same initial BMI category who did not lose weight.[91] A review of eleven population studies, seven from the United States and four from Europe, found the highest mortality rates among those who lost weight or gained a large amount of weight and the lowest mortality rates among those with modest weight gains.[92] Another review study reported conflicting results: two studies found an association between intentional weight loss and decreased mortality, three found an association with increased mortality, and four found no association.[93]

What all of this means is that the resurgence of weight-loss dieting in the 1980s can't have been the result of people newly discovering that fatness is unhealthy and dieting makes you thinner. There was certainly no evidence to support those beliefs in the Gilded Age when weight-loss dieting first became popular. And the vast majority of the research published since that time shows that overweight people live longer than thinner people and dieting almost always fails to produce long-term weight loss. Instead of reflecting some kind of new nutritional wisdom, the sudden concern about obesity, the popular embrace of a meritocratic ideology of thinness, and the proliferation of weight-loss diets seems more likely to have been driven by the same anxiety that drove people to seek status through sophisticated foods. The idea that you could and should be thin by eating right and exercising offered people another alternative arena of aspiration in an era of increasing income inequality and stagnating mobility.

THE FALLACIES IN THE LOCAL AND ORGANIC ORTHODOXY

Compared to the beliefs that gourmet food tastes better and people can and should lose weight by dieting, justifications for the superiority of natural foods are somewhat more complex. People who shop at Whole Foods Market and subscribe to community-supported agriculture (CSA) membership programs, where people pay a farm at the beginning of the growing season to receive a weekly share of whatever they grow, often claim that the food is fresher and better tasting.[94] Locally grown foods are also supposed to be better for the environment, primarily because they do not have to be transported as far. People assume most that local production happens on a smaller scale and in a more transparent way, and these things are thought to promote ecologically and socially responsible choices. Organic produce is supposed to be better for the environment and consumers' health because it cannot rely on synthetic fertilizers and pesticides. Free-range poultry and grass-fed beef are supposed to be more humane because the animals are less confined, better for the environment because their waste isn't as concentrated, and healthier because the animals are supposed to be less susceptible to disease and have a better fatty acid profile.

While this is not a comprehensive list, these beliefs represent some of the core tenets of the contemporary ideal of purity. Many of them were popularized by the 2008 Academy Award–nominated documentary *Food, Inc.* It emphasized the dangers of the industrial food system in interviews with Barbara Kowalcyk, whose two-and-a-half-year-old son Kevin died from an *E. coli* 0157:H7 infection he got from eating a contaminated hamburger. Michael Pollan, one of the documentary's primary consultants and stars, claims that *E. coli* 0157:H7 evolved only after farmers started feeding corn to cattle, which makes their stomachs more acidic. Furthermore, he says that feedlot conditions contribute to contamination because "the animals stand ankle deep in their manure all day long." The implication is that grass-fed beef from free-range cattle is safer. The documentary ends by invoking the oft-cited statistic that the average meal travels 1,500 miles from farm to super-market and offering a series of recommendations for consumers who wish to "vote with their forks" for a better food system: "Visit your local farmer's market. . . . Make it a point to know where your food comes from. . . . Buy organic or sustainable foods with little to no pesticide use."[95] Many consumers and institutions have sought to follow these recommendations at least some of the time, and many more seem to believe that doing so would have myriad aesthetic, nutritional, and ecological benefits.

Proximity Is No Proxy for Sustainability

At the University of Michigan, where I teach, the office of campus sustain-ability set a goal in 2012 that by 2025 at least 20 percent of the food it pur-chases should come from "local and sustainable sources." Acceptable foods include those grown or processed within 250 miles of the Ann Arbor campus, organic produce, free-range poultry, and grass-fed or pasture-raised meat.[96] In the document explaining these guidelines, the section on purchasing local foods says, "While not always the case, the sustainability of food generally increases as the distance it travels from the point of harvest to consumption decreases. Minimizing transportation and refrigeration generally reduces fossil fuel consumption and carbon dioxide emissions."[97] However, it's not clear that this is even "generally" true. Fossil fuel consumption and carbon dioxide emissions depend not only on the total distance covered but also on the fuel efficiency of the transportation method and the weight of the cargo. If a local farmer travels sixty miles to a farmer's market in a pickup truck that gets twenty miles per gallon and carries 500 pounds of produce, a full load of goods moves at about 167 pounds per gallon (or 83 pounds per gallon if you count the trip back to the farm).[98] In the conventional system, food travels an average of just over 1,000 miles from farm to market in a long-haul truck that can carry 40,000–45,000 pounds and averages between 5.5 and

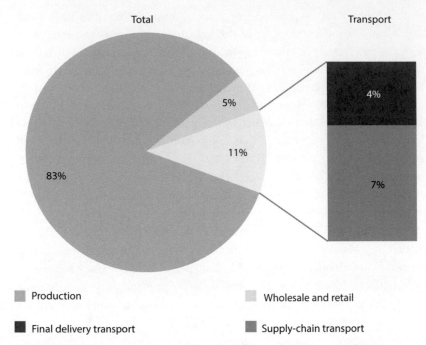

Figure 9. Greenhouse gas emissions for stages of food production and delivery to market. Source: Lindsay Wilson, "The Tricky Truth about Food Miles," *shrink-thatfootprint.com* (2013). Based on data from Christopher L. Weber and H. Scott Matthews, "Food Miles and the Relative Climate Impacts of Food Choices in the United States," *Environmental Science and Technology* 42, no. 10 (2008): 3508–3513. Reproduced by permission.

6.5 miles per gallon.[99] Thus, a full load moves at about 220 pounds per gallon (or 110 pounds per gallon if the truck returns to its origin empty).[100] Rail and container shipping are considerably more efficient than trucking, so food transported by those methods generates fewer emissions per pound even if it travels much further.[101]

Furthermore, transportation—particularly the leg from the farm or factory to the final point of sale or what is called final delivery transport, which is the only part that buying local could even theoretically reduce—accounts for only a small fraction of the energy use and emissions generated by the food system. According to a 2008 analysis of the greenhouse gasses produced throughout the total life cycle of food production, transportation as a whole is responsible for only 11 percent of total emissions (see Fig. 9). Of that 11 percent, 7 percent consists of supply-chain transport. Having many small, widely dispersed farms and factories actually increases the need for supply-chain transport because it increases the total number of deliveries that must be made. Even if buying local totally were to totally eliminate final delivery transport, which it doesn't, it could reduce the emissions generated by the food system by at most 4 percent.[102]

Eighty-three percent of food system emissions come from production. Small-scale or local production often requires more energy and generates more emissions. Studies that account for emissions generated across the entire life cycle of food production frequently conclude that products transported long distances from places where they grow well are responsible for fewer greenhouse gasses than locally grown or made ones. Tomatoes grown in the United Kingdom and transported to London generate 2,394 kilograms of carbon dioxide per ton, while tomatoes grown in Spain generate only 630 kilograms of carbon dioxide per ton, largely because the former must be grown in greenhouses.[103] Similarly, for London consumers, lamb raised in New Zealand and transported by ship is four times more energy efficient than lamb raised in the UK because the former graze outside while the latter are fed grain and spend at least part of the year in heated barns.[104] One of the reasons some areas are known for growing particular crops is because different soil, climate, and weather patterns give them a competitive advantage. Potatoes grow well in Idaho because of the rich volcanic soils, warm days, and cool nights. According to the USDA, in 2013 potatoes yielded an average of 415 hundredweight per acre in Idaho, compared to 240 in North Carolina and Florida, 260 in Massachusetts, and 280 in Arizona.[105] Producing more food on less land means less carbon released by tilling, less pesticide and fertilizer use, less fuel used by farm equipment, less water used in irrigation and lost to evaporation, and more land preserved as wilderness. Wilderness is vastly better at sequestering carbon and fostering biodiversity than even the most diverse farm.[106]

Focusing primarily on how far food travels from the site of production to the point of sale also neglects offsetting factors on the consumer side. Shopping at farmer's markets often requires additional car trips because there are many items people typically buy while shopping for food at grocery stores, such as paper goods and cleaning supplies, that farmer's markets do not sell.[107] CSAs are particularly likely to increase food waste because members have little control over the kind and amount of food they receive each week. A survey of around 200 people who had purchased CSA shares in Iowa found a high rate of attrition. Respondents who were no longer participating gave the following reasons for their withdrawal: there was sometimes too much produce (35 percent), they didn't have time to prepare the food (23 percent), or didn't know how to prepare the food (12 percent).[108]

Some people claim that local production is more sustainable because it is more transparent or promotes the local economy. For example, *Slate*'s Green Lantern environmental advice column told a reader who asked whether subscribing to a CSA was still a good choice if he was throwing much of the produce away, "All else being equal, the more you know about how something is produced, the more likely it is to be environmentally friendly."[109] But even if consumers actually visit the local farms they buy from, how many can evaluate the relative sustainability of the operation? Many crucial factors in evaluating ecological impacts such as greenhouse gas emissions, soil carbon sequestration, and groundwater contamination can't be seen. As rural sociologist Claire Hinrichs writes, "small scale, 'local' farmers are not inherently better environmental stewards."[110] Nor are they necessarily better employers, as Margaret Gray documents in her ethnography of small-scale farms in the Hudson Valley. Just like many of the large farms that supply conventional grocery stores, small farms often rely on poorly paid and systematically mistreated migrant workers.[111] Additionally, while some of the money that small-scale farmers earn may be spent locally, they typically still have to purchase things such as seed, feed, fertilizer, and equipment from distant retailers and service debt held outside the region. The idea that all the money spent on local products will remain within the community is a fantasy.[112] Simply knowing where your food is grown tells you essentially nothing about the ecological, ethical, or economic consequences of purchasing it.

Organic Is Not Safer, Healthier, or More Sustainable

The belief that organic produce is better than conventional often stems from a basic misunderstanding about what organic certification means. Many people, perhaps influenced by films such as *Food, Inc.*, think organic produce is grown without the use of any pesticides or chemicals. While many organic farmers use nonchemical methods of pest control, for example, importing pest predators, the USDA organic certification standards established in 2002 prohibit only the use of synthetic pesticides. They also restrict the use of genetically modified seeds and synthetic fertilizers. Chemicals derived from natural sources such as plants, bacteria, and minerals that do not require extensive processing are approved for use in organic-certified agriculture.

Whether a pesticide or fertilizer is synthetic or natural has no bearing on its toxicity, impact on soil and water quality, or effectiveness in protecting crops. In fact, many pesticides approved for organic use are more dangerous and less effective than the synthetic versions that conventional farmers use. For example, organic fungicides are usually based on copper or sulfur. According to Julie Guthman, sulfur "is said to cause more worker injuries in California than any other agricultural input"[113] and poses a "greater environmental risk than many synthetic fungicides."[114] The organic fungicide copper sulfate is highly toxic to fish, persists indefinitely in the soil where it can kill bacteria and earthworms, has been shown to damage the DNA of mice and reduce the ability of plants to undergo photosynthesis, and causes chronic respiratory problems in workers with prolonged exposure to it. The process of preparing copper is also highly energy intensive, which is one of the reasons it is more expensive than synthetics.[115]

Another organic pesticide, rotenone, was considered safe for decades primarily because it is derived from the roots and stems of a handful of subtropical plants and thus is natural. However, its use was banned in the United States in 2005 because of research showing that it can cause Parkinson's-like symptoms in rats and kill mammals, including humans, in large enough amounts.[116] It remains off limits in the EU, but in 2010, it was reapproved for use in organic farming in America.[117] The combination of rotenone and pyrethrin, another plant-derived pesticide that the EPA classifies as a "likely human carcinogen," is the most commonly used natural insecticide on the market.[118] Not only is that combination more toxic to beneficial insects, fish, birds, and humans than many synthetic alternatives, it is also less effective at killing the target pests. A study conducted in New York apple orchards found that it may require up to seven applications of rotenone-pyrethrin to achieve the same protection as two applications of the less toxic synthetic imidan.[119] Despite the use of sometimes more toxic pesticides in organic agriculture, Guthman says that "there is no question in my mind that, as a rule, organic producers are exposing farmworkers, neighbors, and eaters to far less toxicity than their conventional counterparts." However, the current regulatory framework prevents organic certification from being a guarantee that the farmers in this category will produce more ecologically benign and healthy products.[120]

The USDA Pesticide Data Program does not monitor natural pesticides such as *Bacillus thuringiensis* (Bt) and copper salts.[121] But even though synthetic pesticides are banned from organic farms, a USDA survey conducted in 2010–2011 found residues on 43 percent of organic produce samples.[122] Fortunately, what really matters for the health of the consumer isn't the presence of pesticide residues, natural or synthetic, but whether those residues have any harmful health effects in the amounts that are typically present on food. On that issue, there is a strong consensus that the amount of pesticides ingested by eating produce—whether organic or conventional—poses no threat to consumers' health. According to the EPA, many chemicals that occur naturally in foods people eat every day, such as coffee, basil, mushrooms, lettuce, and orange juice, are more toxic than most pesticides.[123] According to a 1992 *Science* article, the possible human health hazard posed by the known rodent carcinogens in a single cup of coffee (according to the human exposure/rodent potency index) is many times greater than the hazard posed by all synthetic pesticide residues the average American consumer will eat in a year.[124] Ninety-seven percent of the pesticides used in California in 2010 are rated less toxic by the EPA than caffeine or aspirin, and 55 percent are less toxic than vitamin C.[125]

Some people also believe that organic foods contain more beneficial nutrients, but most research has shown no consistent advantages in organic produce itself or to the people who eat more of it. A 2012 review study concluded that neither organic nor conventional produce consistently had higher nutrient levels, and "the published literature lacks strong evidence that organic foods are significantly more nutritious than conventional foods." The same review found that studies on children or adults habitually eating organic had no clinically meaningful differences in any biomarker, nutrient, or contaminant levels.[126]

Because of the restrictions on the use of synthetic fertilizers, some organic farming operations rely on fish meal or guano to replace essential soil nutrients such as nitrogen, phosphorus, and potassium. Much of the guano used in agriculture is harvested on islands off the coast of Peru, where the dry climate preserves the droppings of seabirds, and is shipped to farmers in Europe and North America at great economic and ecological cost.[127] However, most organic farmers use composted manure. Compared to synthetic fertilizers, composted manure has many disadvantages: it is a less balanced nutrient source, it has to be applied in much larger volumes, it may lead to greater nutrient leaching and runoff, it contains traces of antibiotics and heavy metals that may concentrate in the soil, and it generates more greenhouse gas emissions in production.[128] Plant pathologist and agricultural technology consultant Steve Savage calculates that the total carbon footprint of manufacturing the amount of compost typically used on one acre of organic crops is equivalent to the carbon footprint of manufacturing the amount of synthetic urea-nitrogen fertilizer typically used on 12.9 acres of conventional corn. The difference would be even larger if he had included the fuel required to transport and spread the compost.[129]

Perhaps the most important difference between organic and conventional systems in terms of sustainability is the problem of yield. There is an ongoing, often heated, debate about whether and under what circumstances organic methods might produce yields equivalent to or even greater than conventional methods. For some crops, particularly in the developing world[130] and on research test plots,[131] organic production methods can match or even exceed conventional methods. However, most studies so far have concluded that in commercial practice, conventional yields are considerably higher, particularly for major cereal crops such as corn and wheat. According to a 2012 review of sixty-six studies, organic yields are 25 percent lower on average, ranging from only 5 percent lower for crops such as rain-fed legumes to 34 percent lower "when the conventional and organic systems are most comparable."[132] According to USDA data on actual farm production, yields for organic fruits, nuts, and berries are 35 percent lower than conventional, yields for organic field crops such as grains are 30 percent lower, yields for organic vegetables are 38 percent lower, and organic dairies produce 30 percent less milk per cow.[133] Much like shifting to more local production, increasing organic production would likely require using much more land for agriculture, with all the attendant environmental costs.

The Questionable Benefits and Ecological Costs of Ethical Animal Products

Aside from organic certification, most of the alternative certification labels that have appeared in the last few decades relate to the treatment of animals used in agriculture. "Natural eggs" first appeared in the 1980s. They were joined by organic, cage-free, and free-range eggs in the 1990s and omega-3 enriched and pasture-raised eggs in the early 2000s. The proliferation of specialty eggs has left many consumers confused about the differences. According to some surveys, as many as 50 percent of Americans seek out specialty eggs, assuming that labels such as cage-free mean that the hens roam freely on open pasture and produce healthier, better-tasting eggs with less harmful effects on the environment.[134] However, as the University of Michigan Sustainable Food Purchasing Guidelines stipulate, "cage free means that poultry is not always kept in battery cages, but is also not allowed outdoors."[135] Thus, in order to count toward the sustainable purchases goal, eggs must be free range or organic.

However, if purchasing eggs from chickens with outdoor access is the goal, these guidelines may still fall short of their aim. Most free-range and organic eggs are laid by hens that have only token access to the outdoors, usually in the form of a small concrete patio attached to the side of a cage-free coop.[136] Additionally, the industrial-strength fans used to suck ammonia fumes out of the coops often turn the access doors into a vortex of heavy winds, effectively discouraging the chickens from even trying to wander outside.[137] According to a 2010 report by the Cornucopia Institute, 80 percent of USDA organic eggs are laid by hens in industrial-scale coops that typically house 10,000 birds or more in hen houses with no "meaningful outdoor access."[138] According to the American Veterinary Medical Association, cage-free and other alternative systems "also have considerable liabilities in terms of animal health, biosecurity, and economic efficiency." One of those liabilities is nearly double the risk of premature death from fighting and cannibalism. The AVMA advises that "it cannot be assumed that hens in non-cage systems will experience improved welfare."[139] There are costs to giving chickens more meaningful outdoor access, too, in particular a heightened risk of predation.

Many people believe that specialty eggs are healthier, sometimes based on the difference in yolk color that also makes people think they taste better. However, yolk color depends primarily on the concentration of carotenoids in the chicken feed that have no necessary relationship to the nutritional content of the eggs. Some carotenoids, such as beta-carotene, have nutritional value, but not all of them do. According to Marion Nestle, a professor of nutrition, food studies, and public health at New York University, "deeper-colored egg yolks only indicate the presence of carotenoids in general, not necessarily the presence of beta-carotene."[140] Even when the carotenoids are bioactive, there's little reason to believe that the small amount that might be responsible for the difference in yolk color would have a meaningful effect on anyone's health. A typical 50-gram egg usually contains approximately 10 to 15 micrograms of beta-carotene. Pastured eggs may contain as much as 40 to 80 micrograms, but that's still a tiny fraction of the recommended daily intake of 1,500 micrograms or the 3,000 to 4,000 micrograms in a 50-gram portion of carrots or spinach.[141]

Fatty acid composition also plays a role in the debate about the nutritional quality of eggs. However, in 2002, a team of animal scientists at Oregon State University found that specialty eggs were not nutritionally superior to conventional eggs based on fourteen aspects of fatty acid composition, with the exception of eggs specifically labeled omega-3 enriched.[142] The authors explain that while fatty acid composition can sometimes be affected by chicken feed, it also varies based on the age, size, and breed of the hen. Several studies have shown that the eggs laid by chickens raised on pastures with a lot of legumes such as alfalfa and clover may have higher levels of omega-3 fatty acids and vitamins A and E.[143] However, not all grasses provide the same benefits, according to Heather Karsten, assistant professor of crop production and ecology at Penn State. "The leafier the plant, the higher the digestibility for the animal," says Karsten. "If a pasture is overgrazed or the grass is too mature and 'stemmy,' the nutritional benefits fall off."[144] In any case, most specialty eggs come from hens that eat chicken feed composed of corn and soy, just like conventional chickens. And while organic eggs must come from hens that eat organic corn and soy feed, the resulting fatty acid profile is the same.[145]

Even if specialty egg-laying operations aren't any more humane and there aren't any clear health benefits for consumers, is it possible that they're better for the environment? This turns out to be a difficult question to answer. If the chicken feed is grown using less effective and potentially more toxic pesticides and less effective fertilizers that result in more greenhouse gas emissions and produces lower yields than conventional methods, the resulting eggs are not likely to be more sustainable. Cage-free, free-range, and pastured egg operations also require significantly more land than battery-cage ones because the cages can be stacked. In some diverse farm systems that raise chickens on pasture, the chicken manure becomes a source of nitrogen enrichment that may improve complementary crop yields. However, a typical free-range, cage-free, or organic egg-laying system makes chicken waste harder to collect and control than in the typical battery-cage hen house, resulting in higher emissions of ammonia, an environmental toxin that primarily affects the workers and the people who live nearby.[146]

There is one factor where the ecological comparison between different egg production systems is clear, and that is feed efficiency. According to data from the European Union, chickens raised in battery cages take about two kilograms of feed to produce one kilogram of eggs. Cage-free chickens with no meaningful outdoor access require 14 percent more food, free-range birds who have access to the outdoors require 18 percent more, and organic chickens need 20 percent more.[147] The main reasons for the difference, according to Hongwei Xi, director of the Egg Industry Center at Iowa State University, are that uncaged birds walk around a lot more, expending more energy than caged birds. The cage-free systems have a slight advantage over free-range ones because it's easier to control the temperature in fully enclosed coops, so the chickens burn fewer calories keeping themselves warm. There are other factors working against the specialty eggs, too. Cage-free hens have higher rates of mortality and injury caused by fighting and cannibalism, and their eggs have thinner shells that are significantly more prone to breakage.[148] Ultimately, if your top priority is chicken welfare and you're willing to pay more to promote it, eggs from pasture-raised hens are probably the best choice. But if you are more concerned about the environment, conventional eggs from battery cage hens probably have the smallest footprint.

The picture is similarly complicated when it comes to free-range, grass-fed beef versus conventional, grain-finished beef. First, it's worth noting that all beef cattle spend the majority of their lives roaming freely on pasture. For the first six to twelve months of their lives, their primary source of sustenance is milk from their mothers. Then they spend another twelve to eighteen months eating grass, forage, and hay. It's only after that point that the lives of grass-fed and conventional beef cows diverge; the latter are transferred to concentrated animal feeding operations (CAFOs), where they eat a grain- and soy-based feed in confinement for the final sixty to ninety days of their lives. Grass-finished cattle continue to eat grass or hay until their deaths and usually take an additional two to four months to reach "slaughter weight." By most accounts, the pasture is far more pleasant than the CAFO, so in terms of welfare, grass-finished cows fare better in the final months of their lives.

Several studies have found that the beef from grass-finished cows has less saturated fat and more omega-3 fatty acids and conjugated linoleic acid, all of which are presumed to be beneficial.[149] However, a large meta-analysis on the effects of consuming different amounts of these fatty acids found essentially no difference in risk of cardiovascular disease between people who eat the most saturated fat and those who ate the least.[150] The association between omega-3 fatty acids and cardiovascular disease is the subject of ongoing investigation and considerable controversy. A 2012 review and meta-analysis of twenty randomized controlled trials found no significant relationship between omega-3 consumption and cardiovascular disease or death and described "major diversity" and inconsistency in the studies included.[151] This inconsistency is exemplified by the conflicting results of two studies on omega-3 consumption in food (instead of from supplements) led by the same researcher. The 1989 study found that men recovering from a heart attack who were advised to increase their consumption of fatty fish were less likely to die in the two years during which they were followed than the men who did not receive the same advice (a relative risk of .73 for the men advised to eat fatty fish compared to the control); the 2003 study found that men with angina who were advised to eat more oily fish were more likely to suffer from cardiac death in the 3–9 years during which they were followed than men who did not get the same advice (a relative risk of 1.27 for the men advised to eat fatty fish compared to the control).[152] It is unclear but doubtful whether eating exclusively grass-finished beef instead of corn-finished beef would have any meaningful cardiac health benefits.

The claim in *Food, Inc.* that feeding grain to cows caused the emergence of more dangerous food-borne pathogens such as *E. coli* 0157:H7 seems to have originated with a 2006 *New York Times* op-ed by Nina Planck, a farmers' market entrepreneur and the author of several books advocating real-food diets. Planck based the claim on a study involving just three cows in which the grass-fed cow had fewer total *E. coli* in its manure than a corn-fed one.[153] However, not all strains of *E. coli* cause disease, and larger studies have found that the pathogenic strain *E. coli* 0157:H7 is no more common in grain-fed cattle than grass-fed cattle. Additionally, several outbreaks have been traced to free-range cows on pasture.[154] It is true that cattle that are fed corn instead of hay have more acidic stomachs, and some bacteria do seem to develop resistance to acid shock in corn-fed cattle, which means that they would be more likely to survive in the human digestive tract. But studies designed to test whether diet influences the prevalence or acid resistance of *E. coli* 0157.H7 have concluded that it has no effect, likely because that particular strain colonizes the section of the gastrointestinal tract farthest from the rumen, meaning it is not affected as much by the change in stomach pH.[155]

Ecological differences are even less clear for beef. Finishing cattle on grass requires more land for pasture, but that may be offset by the land required to produce the grain and soy used in feed. The ecological impacts of both grazing and growing feed depend on many factors. Well-managed pasture needs little or no fertilizers or pesticides and may be better at sequestering carbon, preventing erosion, and fostering biodiversity than land cultivated for crops. In contrast, overgrazing can promote erosion and desertification.[156] Additionally, grass-finished cattle yield significantly less beef per carcass, which increases the environmental impact per pound. In one Australian study, the per-head carbon footprint of grain-finished cattle was higher but the per-pound footprint was lower.[157] Again, if your primary concern is the humane treatment of agricultural animals, grass-finished beef cattle probably suffer less than cows in CAFOs. Ecologically, neither is the clear winner. Overall impacts depend on the particularities of individual operations.

This information isn't hidden away in arcane or subscription-only scholarly journals. A plethora of mass media articles and popular press books have called attention to the logical and factual errors in the standard recommendation to buy local, organic, free-range, grass-fed, and so on.[158] At least one international gathering of climate scientists has rated the "planet saving ability" of the local food movement "poor."[159] Book-length investigations of the organic and local food movements, in some cases by impassioned advocates of sustainability and social justice, have challenged the prevailing faith in the superiority of foods marketed as natural.[160] And yet these arguments have so far failed to resonate with a wider audience. Perhaps the reason is that the popularity of natural food is no more based on an enlightened, rational response to the available evidence than the elevation of gourmet food and thinness are. People believe in the superiority of natural food because they want to believe in its superiority. It provides them with way of performing and embodying a higher social status that they can convince themselves is virtuous instead of pretentious.

THE MISGUIDED PURSUIT OF AUTHENTICITY AND EXOTICISM

The ideal of cosmopolitanism is primarily justified by the belief that there is such a thing as authentic food and that authenticity is a very good thing. Johnston and Baumann's 2010 book *Foodies*, featuring the timeline of gourmet culture that begins in the 1940s, argues that "authenticity is a key element of how foodies evaluate and legitimate food choices" and says that the idea of authenticity in food commingled with the "relentless search for exotic new ingredients, new cuisines, and undiscovered foods."[161] The authors note that perceived ethnic connections are only one of many factors that can signal authenticity in food. Others include simplicity, a personal connection, geographic specificity, and historical tradition. However, most of the other factors work as proxies for ethnicity. Oaxacan mole may also refer to a particular location, pit barbeque has a long history, and sushi prepared by a Japanese chef provides a personal connection, but all of those factors primarily serve to ratify the distinctive otherness of the food.

Signs of authenticity are rarely straightforward. Many foods seen as simple, such as hand-made tortillas or ricotta, are more difficult than factory-made versions to procure and are generally seen as aesthetically more complex. The belief that they are simple and thus authentic derives not from any inherent qualities, such as the number of ingredients they contain, but from their seeming distance from modern industrial life. In many cases, ethnic food gives off an aura of authenticity because of lingering beliefs that non-Western people are more primitive.[162] Whereas mass-produced ketchup and soft drinks generally seem placeless and modern, ethnic and foreign foods seem geographically rooted and traditional. Both of these assumptions are false: all foods come from somewhere and all foods have a history. Furthermore, the origin stories of many authentic foods are more mythical than factual. But the primary flaw in the search for the authentic isn't that people get their histories wrong so much as the instability of the concepts themselves. There is no such thing as an inherently authentic food, and the essential arbitrariness of the social construction of the ideal undermines the notion that its rising cultural value is the result of some kind of enlightenment.

Many foods are composed of few ingredients and are traceable to one or more specific locations and historical traditions and to ethnic groups that don't pass muster as authentic. Consider Lay's Classic potato chips. Potato chips are sometimes traced to Saratoga Springs, New York, based on a popular legend that they were invented there in 1853 by George Crum, the black and Native American cook at Moon's Lake House, in response to a customer complaining his fried potatoes were cut too thick. However, recipes for potatoes shaved thin and fried crisp appear in earlier cookbooks too, such as William Kitchiner's *The Cook's Oracle* (1822) and Mary Randolph's *The Virginia House-Wife* (1824). The mass production of potato chips began over a century ago with the founding of companies such as New England's Tri-Sum Potato Chips (originally the Leominster Potato Chip Company) in 1908 and Mike-sell's Potato Chip Company of Dayton, Ohio, in 1910. These early manufacturers distributed the chips by horse and wagon to merchants who stored them in storefront glass bins or barrels and scooped them into stapled waxed-paper bags to order. By 1932, when Herman Lay founded his snack food company in Nashville, the widespread adoption of cellophane packaging had enabled manufacturers to bag foods such as potato chips at the factory. Today, his company, which merged with the corn chip manufacturer Fritos in 1971, boasts that Lay's Classic potato chips have been "America's favorite snack for nearly 75 years."[163] They contain only three ingredients: potatoes, vegetable oil, and salt. But none of that is sufficient to make Lay's potato chips authentic in most people's eyes.

The problem cannot be simply that Lay's is an industrial product. Mass-produced Sriracha (aka "rooster sauce") made by Huy Fong Foods is widely seen as an authentic Vietnamese condiment despite the fact that its origins are no more exotic than potato chips and are much more recent. Sriracha was invented in Los Angeles in 1983 by David Tran, an immigrant from a Cantonese Chinese enclave in Saigon. Dissatisfied with the hot sauce options available in the United States, he began making what he called "Pepper Sa-Té Sauce" in 1980. A few years later, he hit on the blend he called Sriracha, which quickly became a huge success. The official name comes from the southern coastal town in Thailand where a similar condiment is supposed to have originated, and the nickname comes from the rooster on the bottle, a nod to Tran's sign in the Chinese zodiac. What makes Sriracha different from most Mexican and Louisiana-style hot sauces is ordinary garlic powder. The other ingredients are jalapeños, sugar, salt, vinegar, potassium sorbate, sodium bisulfate, and xanthan gum—not nearly as simple as Lay's.[164] The condiment's California origins are advertised on the front of every bottle, but it nonetheless maintains an exotic aura, which is probably responsible for at least some of its popular cachet.

The key difference between Sriracha and Lay's is that the former seems like an ethnic food, a slippery category based on the assumption of an ethnically unmarked, white vantage point. Although both Sriracha and the potato chip are American inventions, the spicy condiment with a Thai name that is made by an ethnically Chinese immigrant and is strongly associated with Vietnamese cuisine is distanced from the mainstream in a way that a snack that could be traced to Euro-American figures such as Mary Randolph and Herman W. Lay, or even the biracial George Crum, is not. But fried potatoes have a foreign pedigree of their own, too.

Potatoes were originally cultivated in the Andes and made their way to Europe during the fifteenth and sixteenth centuries during the Columbian Exchange, the great transfer of animals, plants, people, ideas, and technologies between the Americas, Africa, and Europe. Smaller and more bitter than most varieties grown today, potatoes were initially scorned by most Europeans. They were seen as similar to turnips, which were commonly used as winter cattle feed. However, the ruling classes in countries from France to Finland to Russia began to encourage the rural poor to eat them, especially during the famines and bread shortages of the eighteenth century. People gradually bred more palatable varietals and invented novel culinary applications, among them fried potato dishes such as the latke, which is associated with Eastern European Jewish communities, and the French fry, which is typically traced to Belgium. Recipes in *The Virginia Housewife* and the menu at Moon's Lake House are typical of the cuisine the emerging middle class developed in Britain in the eighteenth century.[165] So the potato chip is no less a product of global cultural flows than Sriracha, but Belgium and Britain seem less foreign to most Americans than Thailand and Vietnam.

There is no single universal reference point for otherness. In a playful nod to the slippery nature of exoticism, the Original Chinatown Ice Cream Factory in New York City separates their menu into a list of "Regular Flavors" that includes almond cookie, black sesame, durian, ginger, green tea, lychee, mango, pandan, red bean, and taro/ube and a list of "Exotic Flavors" that includes chocolate, mint chip, Oreo cookie, pistachio, rocky road, strawberry, and vanilla. However, the dominant construction of the exotic typically presumes a Euro-American perspective as the norm. British cuisine in particular is often seen as basically identical to American cuisine.

Nevertheless, many quintessentially British foodways are far more foreign to most Americans than tacos and sushi. Formal afternoon tea service with scones and clotted cream and Anglo-Indian dishes such as kedgeree and chicken tikka masala still seem exotic. Euro-Americans whose ancestors settled in the United States generations ago often retain food traditions from the "Old World" that remain largely unknown outside their ethnic community or regions where large numbers of immigrants from that community settled.

Annual festivals celebrate the Norwegian flatbread lefse in Minnesota and North Dakota, the Polish community centered near Hamtramck, Michigan, has made the jam-filled donuts called paczki a Fat Tuesday tradition throughout greater Detroit, and the Danish-American kringle was named the official state pastry of Wisconsin in 2013. These regional European ethnic specialties are foreign to most Americans even as many other foods once associated with more marginalized ethnic groups, such as spaghetti, pickles, burritos, and chop suey, have been thoroughly assimilated into the mainstream.

It may seem as if there is no consistent logic or pattern to what kinds of foods are deemed exotic and which ones are unmarked. However, the construction of the authentic and exotic can be mapped on a kind of continuum that reveals a telling implicit logic. A local, non-chain restaurant owned and operated by an Italian American family will typically be seen as more authentic than Olive Garden, which in turn seems more authentic than a can of Chef Boyardee spaghetti or ravioli. However, Italian food consumed at a restaurant in Italy is even more authentic, and the less that restaurant is seen as catering to tourists, the better. Maximum authenticity points would likely be awarded to food prepared by an Italian grandmother in her rural Italian home, using ingredients harvested from her kitchen garden.

The distinctions as you step up from Chef Boyardee to the Italian grandmother don't necessarily correspond to simplicity, location, or history, but they do reflect increasing barriers to access. The can of Chef Boyardee[166] is accessible and affordable to almost anyone. Each subsequent rung on the authenticity ladder requires more money, knowledge, mobility, or social capital to reach. What really makes something authentic, then, has less to do with any actual connection to a particular place, time, or group of people than it does with the kind of scarcity that makes particular goods valuable in status competitions. The connections can be invented to suit the relative accessibility.

In *The Authenticity Hoax*, Andrew Potter argues that authenticity is always a positional good, not just when it comes to food:

> Being spontaneous, living in a small community, shopping locally—all of these are subject to an indefinite number of divisions and gradations. When it comes to shopping locally, how local is local enough? How risky is risky enough? If we want to have a low-impact, environmentally conscious lifestyle, how far do we need to go? Being an authentic person, or living an authentic life, turns out to be not so different from being a nonconformist: it is a positional good that derives its value from the force of invidious comparison. You can only be a truly authentic person as long as most of the people around you are not.[167]

Potter attributes the obsessive and futile search for authenticity to the disenchantment the three major developments that heralded modernity—capitalism, science, and liberalism—ushered in. He claims that modern malaise encourages people to imagine a lost wholeness they seek to recover. The predatory character of class conflict turns the search for wholeness into a competition. However, this theory doesn't quite explain why the interest in ethnic and international food would wax and wane the way it has over the twentieth century. The influence of shifts in class structure on the nature of class conflict provides the missing piece. It makes sense that changes in inequality and mobility would affect how motivated people are to participate in the zero-sum competition for status. As with the other three ideals of the food revolution, the growing interest in and anxiety about authentic and exotic food in the 1980s could have been driven by shifts in class structure that created larger incentives for people to seek symbolic distinction.

The belief that these ideals were embraced because of a collective culinary enlightenment allows those who invest most deeply in them to flatter themselves that their tastes are the result of superior knowledge, refinement, and virtue. Although the assumptions these ideals are based on—that gourmet foods taste better to anyone with a well-trained palate, that thinness is meritocratic, that natural foods are better for the environment, and that authenticity is real—reflected much the messages of contemporary mass media about food, that doesn't mean that the people who embrace these beliefs have been brainwashed. In fact, media representations of food are often deeply ambivalent, reflecting many of the tensions within and between the different ideals. However, audiences interpret media in ways that serve their interests. The next three chapters explore this process, looking at how the popular discourse about food sometimes promotes the ideology of the food revolution and how audiences make sense of conflicting media representations and use them to negotiate their place in the social hierarchy.

CHAPTER 4

Anyone Can Cook

SAYING YES TO MERITOCRACY

Were we to evaluate people, not only according to their intelligence and education, their occupation and their power, but according to their kindness and their courage, their imagination and sensitivity, their sympathy and generosity, there could be no classes.

—Michael Young, inventor of the term "meritocracy"

The critically acclaimed 2007 Pixar film *Ratatouille* begins by zooming in on a television set broadcasting the image of a spinning globe with an Eiffel Tower sticking out to mark the location of France. A voice with a French accent says, "Although each of the world's countries would like to dispute this fact, we French know the truth: the best food in the world is made in France. The best food in France is made in Paris. And the best food in Paris, some say, is made by Chef Auguste Gusteau. . . . Chef Gusteau's cookbook, *Anyone Can Cook!* climbed to the top of the bestseller list. But not everyone celebrates its success." Then a montage of images of Paris and Chef Gusteau cuts to an interview with a skinny bald man identified by the screen caption as "Anton Ego, Food Critic, 'The Grim Eater.'" Ego rebalances the half-moon spectacles on top of his giant nose as he says, "Amusing title, *Anyone Can Cook!* What's even more amusing is that Gusteau actually seems to believe it. I, on the other hand, take cooking seriously. And, no, I don't think anyone can do it." He punctuates the statement by tossing the cookbook off screen. In less than a minute, this opening sequence establishes the central conflict in the film: Gusteau's egalitarian, inclusive philosophy versus Ego's elitist, exclusive one.[1]

It will come as little surprise that Ego is the villain. Most Americans are deeply invested in the notion that people should be rewarded in proportion to their talent and effort and that everyone should have the opportunity to succeed. The basic tenets of this ideology, commonly referred to as meritocracy, apply not only to concrete markers of success such as wealth but also to symbolic goods such as status and respect. All four of the ideals of the food revolution are justified ideologically by meritocratic narratives. Appreciating foods that are seen as gourmet, healthy, natural, or ethnic is taken as evidence of being cultured, intelligent, and virtuous. Having good taste is portrayed as an achievement rather than merely an expression of arbitrary preferences or inherited privilege. A counternarrative portrays expensive or rarefied tastes as potentially snobbish; this is the subject of the next chapter. This chapter focuses on the dominant narrative: how mass media texts portray the idea that aspirational eating is both admirable and accessible to anyone and how audiences respond to these stories.

In *Twilight of the Elites*, Christopher Hayes claims that the moral justification of meritocracy is based on a distinction between contingent and essential features: "People are not discriminated against due to contingent, inessential features like skin color, religious affiliation, or gender, but rather due to their essential features: their cognitive abilities and self-discipline." The problem, Hayes explains, is that the line between contingent and essential features is blurry. "What we call intelligence, along with work habits, diligence, social abilities, and a whole host of other attributes we associate with success, seem to emanate from some alchemical mix of genetics, parental modeling, class status, cultural legacies, socioeconomic peers, and early educational opportunities."[2] If most of these characteristics are a combination of nature and nurture, how do we avoid rewarding people for things they have no control over? The issue of essential versus contingent features is magnified when it comes to taste hierarchies because all but the most solipsistic aesthetes recognize that taste is at least somewhat subjective. As the opening to *Ratatouille* acknowledges, people from countries other than France might dispute the "fact" that French food is the best. Tastes, eating habits, and bodies are profoundly shaped by precisely those contingent factors that meritocracies are not supposed to discriminate against, such as race and ethnicity, socioeconomic background, and gender.

Stories about gourmet culture such as *Ratatouille* and weight-loss narratives such as *The Biggest Loser* reveal how good taste and thinness come to seem meritocratic despite the unevenness of the playing field. In the first, a lowly rat overcomes the limitations of his species to become a gourmet cook who can impress even the most discerning critic. In *The Biggest Loser*, fat people compete to see who can lose the most weight through diet and exercise with the help of celebrity personal trainers. Although both seem to endorse the logic of meritocracy, they actually reveal contradictions to the idea that anyone can cook or get thin. Instead of being truly egalitarian,

Ratatouille constructs a new hierarchy of the palate and suggests that Remy's cooking skill is an extraordinary gift, not something just anyone can do. On *The Biggest Loser*, contestants' bodies sometimes defy the expectation that weight is a transparent reflection of eating right and exercising. However, in both cases, audiences interpret the stories as affirmations of the idea that anyone can become a great chef and anyone can become thin in spite of those contradictions.

These examples suggest that faith in meritocracy is not the result of Americans being duped by the media, as some scholars have suggested. For example, in *The New Class Society*, Robert Perrucci and Earl Wysong claim that class inequality is made possible by cultural products that persuade people that the way things work out is fair: "It is precisely because of . . . *Forrest Gump*–genre films, and other 'rags-to-riches' cultural products that most Americans appear to accept the myth of the American Dream."[3] *Ratatouille* and *The Biggest Loser* demonstrate how in many cases, "rags-to-riches" stories are ambivalent about the relationship between merit and rewards or even work to challenge the idea that success is available to anyone who works for it. These success stories may resonate with a large audience precisely because they portray challenges to meritocracy that account for the capriciousness of success. People want to believe that the system of rewards they're invested in are fair, even if their firsthand experiences say otherwise. Viewers actively negotiate with these ambivalent narratives about taste and thinness, often reading against the grain to make the stories mean that they are truly meritocratic.

THE PALATE ELITE

The opening of *Ratatouille* explains that Chef Gusteau was the youngest chef in Paris to earn five stars but that one of the stars was rescinded after Ego published a scathing review. Gusteau was brokenhearted and died shortly thereafter. However, his can-do philosophy lives on in the film's protagonist, a rat named Remy who starts talking to a figment of the chef after he gets separated from his colony and ends up alone in the sewers of Paris with a copy of *Anyone Can Cook!* The orphaned Remy teams up with Linguini, who was recently hired as the garbage boy at Gusteau's restaurant. Linguini has no cooking skill of his own, but through some trial and error, the two of them discover that Remy can control Linguini like a marionette by tugging on his hair while remaining hidden from view by a chef's hat. The pair manages to fool the other chefs at Gusteau's restaurant into thinking that Lingiuni is a culinary genius, and Remy's cooking begins to lure back the Parisian crowds.

Ego becomes incensed that the restaurant is becoming popular again despite the fact that he has given it a negative review and announces that he will return to "deflate the overheated puffery" about the new chef.

On the night of Ego's return visit, Remy takes a risk by preparing rata-touille, traditionally a simple vegetable stew the other chefs call a "peasant dish," not something a five-star restaurant would normally serve. But Remy's version wows Ego, who agrees to wait until closing time for the privilege of meeting the chef. After the last guest has gone, Linguini and the one other chef who knows about Remy introduce him to the critic and explain how a rat came to be the primary architect behind the restaurant's renaissance. The review Ego writes resolves the conflict set up in the opening scene: "In the past I have made no secret of my disdain for Chef Gusteau's famous motto, 'Anyone can cook.' But I realize only now do I truly understand what he meant. Not everyone can become a great artist, but a great artist can come from anywhere." Thus, the film seems to side with Gusteau's philosophy. Even Ego agrees that social origins are irrelevant; a rat can become a great cook, and a peasant dish can become an entrée worthy of a five-star restaurant.

However, the interpretation of Gusteau's motto that Ego endorses is not actually inclusive. Ego claims to have previously misunderstood Gusteau to be suggesting that everyone has the ability to become a great artist. What he agrees with in the end is that the social origins of a great artist don't matter. Gusteau's own explanation of his motto is more nuanced. He says, "Great cooking is not for the faint of heart. You must be imaginative, strong-hearted, you must try things that may not work. And you must not let anyone define your limits because of where you come from. Your only limit is your soul. What I say is true, anyone can cook, but only the fearless can be great."[4] In essence, Gusteau is making a distinction between the contingent and essential features in the meritocracy of the kitchen. Social origins are contingent; creativity, courage, and ambition are essential. The quote implies that other people may try to "define your limits because of where you come from," and Gusteau and Ego both reject that kind of discrimination, but they both also suggest that achieving greatness is reserved for a select few.

Remy does show courage and ambition by repeatedly putting himself in harm's way in order to pursue his desire to create great food. He also demon-strates creativity by defying the other cooks in the kitchen who want to stick to Gusteau's recipes and by insisting repeatedly that he would rather create than steal the way his fellow rats do. However, the film suggests that those factors are ancillary to his ultimate success. The primary reason Remy is driven to create great food is his "highly developed sense of taste and smell." Instead of suggesting that great cooking is truly something anyone can do, the film re-mystifies culinary skill as something you must have innate talent to achieve and perhaps also to appreciate. Ultimately, *Ratatouille* does not advocate a truly inclusive philosophy in which anyone who tries hard can be great; instead, it suggests that culinary sophistication is restricted to those who possess a sensitive palate.

The superiority of Remy's palate is established in two scenes in which taste and smell are represented visually and aurally in a sort of cinematic synesthesia. The first occurs approximately four minutes into the film when Remy sneaks into the kitchen of the house where his colony is living. Gusteau appears on a television that is visible from the kitchen counter, saying, "Good food is like music you can taste, color you can smell. There is excellence all around you. You need only be aware to stop and savor it." Inspired, Remy looks around and finds a bit of cheese and a strawberry. He closes his eyes, and the kitchen disappears so he is shown against a plain black background. First he takes a bite of the cheese. Soft round shapes in golden hues appear to his left while a syncopated brass melody plays. Then he takes a bite of the strawberry. Delicate pink and purple swirls appear to his right and lilting strings replace the brass. Finally, he takes a bite of both at the same time, and the entire screen fills swirling shapes in both orange and pink as the brass and string themes merge harmoniously (see Fig. 10).

The sequence is repeated about halfway through the film as Remy tries to guide his brother Emile through a similar taste experience. In the alleyway behind Gusteau's restaurant, Remy stops Emile from eating an unidentifiable piece of garbage, saying, "I have got to teach you about food!" He tells Emile to close his eyes, and this time it is Emile who is shown against a plain black background. Remy hands him a piece of cheese and Emile swallows it greedily, with crumbs scattering from his mouth, emphasizing the difference

Figure 10. Remy eats strawberries and cheese. Source: *Ratatouille*, dir. Brad Bird, DVD, Walt Disney Pictures, 2007.

between his natural inclinations and his brother's. Remy cries, "No, no, no! Don't just hork it down!" and then, scowling, offers him another piece with the instruction, "Chew it slowly . . . think only of the taste. See?" A vague blob appears next to Emile's head and faint, disjointed brass notes play. Remy tries to nudge him along: "Creamy, salty sweet. An oaky nuttiness?" to which Emile replies sarcastically, "Oh, I'm detecting nuttiness." Undaunted, Remy hands him the strawberry, saying, "Now taste this. Whole different thing, right? Sweet, crisp, slight tang on the finish." Another blob appears in the same place, a little brighter this time. "Okay," Emile allows, tentatively. "Now try them together," Remy instructs. Emile chews carefully, eyes shut tight in concentration. The blobs begin to form small shapes with a little more color, and the notes begin to sound more melodic (see Fig. 11). "Okay," Emile says, "I think I'm getting a little something there. It might be the nuttiness. Could be the tang."

Triumphant, Remy begins to rave about how many different tastes there are in the world and how many combinations must be possible, breaking Emile's concentration. As Emile opens his eyes, the music stops abruptly, the colors disappear, and the alleyway returns. "I think you lost me again," he says, and Remy looks disappointed. The same foods that conjured up vibrant, distinctive sounds, shapes, and colors for Remy are barely differentiated for Emile. For Remy, the different flavors initially appear on opposite sides of him, while for Emile they appear together in the same place. For Remy, the combined flavors fill the screen with color and the soundtrack with complex symphonic music, while for Emile, the colors are faint and the sounds disjointed. The limited progress Emile makes with Remy's coaching may suggest that even with a naturally insensitive palate, he might be able to develop a

Figure 11. Emile eats strawberries and cheese. Source: *Ratatouille*, dir. Brad Bird, DVD, Walt Disney Pictures, 2007.

modest appreciation for fine food. However, the primary message is that Remy's palate is exceptional. As Emile puts it, Remy has a unique gift.

This gift endows Remy with remarkable cooking skills, as is demonstrated by the contrast between him and the other chefs at Gusteau's restaurant. The other chefs are all portrayed as talented, skilled professionals who've risen to the top in a competitive environment. After Linguini is promoted from garbage boy to chef thanks to Remy's puppeteering, he receives some tutoring from the tough-talking, motorcycle-riding chef Colette. She scolds him for cutting vegetables too slowly and then says, "You think cooking is a cute job, like Mommy in the kitchen? Well, Mommy never had to face the dinner rush when the orders come flooding in and every dish is different and none are simple and all of the different cooking times but must arrive on the customer's table at exactly the same time, hot and perfect!" Then, after demonstrating the correct posture and form for chopping vegetables, she introduces some of the other chefs: a former circus acrobat who was fired for messing around with the ringmaster's daughter, a convict, a card shark, and a former gunrunner for a failed resistance movement. "So you see," she declares with flourish, "we are artists, pirates, more than cooks are we!" This scene establishes that cooking at a gourmet restaurant is challenging work and that the chefs in Gusteau's restaurant are there because they excel at it, not because they come from some kind of elite background. They seem to be the very embodiment of Gusteau's motto. However, they do not create. Colette insists that they must stick to Gusteau's recipes, even when Linguini is assigned by the malevolent head chef to make a sweetbread dish with cuttlefish tentacles, snail porridge, Douglas fir purée, and veal stomach, a dish that Gusteau himself admitted was a disaster. Instead of obeying her, Remy sends Linguini scrambling around the kitchen to gather new ingredients and improvises a sauce that he manages to sneak onto the plate just before it leaves the kitchen. The customers declare Remy's version delicious.

Remy's exceptional skill is also represented by the ratatouille. After Remy insists that ratatouille is what he wants to serve Ego despite the fact that it's a peasant dish, Colette shrugs, grabs a sprig of herbs, and turns to add it to a pot. Remy stops her. "What?" she asks, "I'm making the ratatouille." He shakes his head no, and she asks, "Well, how would you prepare it?" Remy shows her step by step, indicating that she should slice the vegetables thinly with a mandoline and bake them under parchment paper. He puts the final touches on the plate himself. For all her toughness and skill, Colette is just a grind whose role in the kitchen consists of executing someone else's vision. Remy is apparently the only real artist in what's supposed to be one of the best restaurants in all of Paris.

At the climactic moment, Remy's ratatouille is delivered to a scowling
Ego. He takes a bite and his eyes grow wide. The restaurant whooshes out of
focus as the scene cuts to a flashback from Ego's childhood: he is pictured as
a young boy standing in the doorway of a small cottage, sniffling and fighting
back tears. A wrecked bicycle is visible on the pathway to the house behind
him. His mother is standing at the stove, and when she turns around and sees
what has happened, she smiles sympathetically. Moments later, the young Ego
is seated at a table in the kitchen, still sniffling. His mother places a steaming
bowl in front of him, filled with a traditional stew-like ratatouille, and touches
his cheek tenderly. He takes a bite, and the scene cuts back to the restaurant.
Ego remains frozen in shock for a moment, and then his face is transformed
by a smile. He devours the rest of the dish with delight. Even the great Chef
Gusteau couldn't earn that kind of reaction from the picky critic.

Despite the seemingly democratic motto "anyone can cook," *Ratatouille*
ultimately reinforces an exclusive taste hierarchy. In order to create a ratatouille suitable for haute cuisine, the filmmakers consulted with Thomas
Keller, the owner and head chef of The French Laundry in Yountville,
California, and Per Se in New York City, both of which have three Michelin
stars, the highest rating awarded by the prestigious international restaurant
guidebooks.[5] Unlike the rustic bowl of stew Ego's mother serves him, the ratatouille Remy creates is a delicate sculpture with layers of color reminiscent
of Bourdieu's description of bourgeois meals: "Dishes considered as much
in terms of shape and colour (like works of art) as of their consumable substance."[6] The film may formally reject the idea that great cooking is based on
elitist criteria, but Remy's success is ultimately ratified by the opinion of an
elitist critic. Ego's reaction to the dish is portrayed as deeply subjective and
rooted in a humble childhood memory, but the reason his opinion matters is
because he is one of the gatekeepers of the official taste hierarchy.

MAKING *RATATOUILLE* MERITOCRATIC

Reviews suggest that the film's message was key to its popular success and critical acclaim. *Ratatouille* was nominated for five Academy Awards and won the
Oscar for Best Animated Feature Film in 2007. It was also the eleventh highest-grossing film of 2007 in the United States, and as of April 2016, it is still
the 154th highest-grossing film of all time.[7] According to Rotten Tomatoes, 96
percent of 239 professional critics' reviews were positive. In addition, 87 percent of the million-plus users who have rated the film gave it a positive rating.[8]
On Metacritic, which translates reviews into numerical grades, *Ratatouille* has
an average grade of 96 from professional critics, which qualifies as "universal

acclaim" and makes it the eighth highest-rated film of all time on the site. Its average grade from 1,296 Metacritic users is 8.6 out of 10.[9] Both professional critics and the users of Metacritic, Rotten Tomatoes, and IMDb praise the film for its inspiring moral or lesson, but there are some telling differences in how those two groups interpret what exactly that lesson is.

Of the thirty-five professional critics whose reviews collectively constitute *Rataoutille*'s stellar grade on Metacritic, five of them (or roughly 14 percent) describe its message as democratic or inclusive. Two (6 percent) describe it as exclusive or elitist. A. O. Scott's review in the *New York Times* did both, declaring the film to be "exuberantly democratic and unabashedly elitist, defending good taste and aesthetic accomplishment not as snobbish entitlements but as universal ideals."[10] Many other critics focused on the theme of artistry. In fact, the two themes that professional critics were most likely to mention in their reviews are the triumph of an outcast or underdog (34 percent) and the importance of art (34 percent). Instead of seeing Remy's quest to become a chef as a parable for everyone's capacity for achievement, the professional critics generally interpreted the film as a celebration of "gifted outsiders"[11] who transcend their humble backgrounds and societal prejudices or as a paean to how "fine art—in this case the kind that comes in a cream sauce, with morels on the side—nourishes the soul."[12] The critics who focused on what the film has to say about art usually suggest that Remy is exceptional and rare. For example, Kyle Smith of the *New York Post* wrote, "Remy's quest for excellence is magnificently done; he stands in for all artists who take risks"; and Scott Foundas of the *Village Voice* said that its plot is a "heady brew about nothing less than the principles of artistic creation . . . the disparity between art and commerce, and between those lives confined to ordinariness and those meant for Olympian heights."[13] Although many of the critics described the underdog story as inspiring and said that children would relate to it, few suggested that Remy's achievements are accessible to everyone.

In contrast, of the 222 reviews on IMDb that mention the film's message,[14] 44 (or 20 percent) specifically described that message as inclusive and applicable to anyone. A typical review began this way: "Remy, a rat with ironically refined tastes in matters culinary, embodies the film's adamantly pro-democratic theme of 'anyone can cook'" and describes the film overall as "a powerful primer for the film's target audience on American democracy and the egalitarian can-do notion at its core."[15] Another IMDb user said, "The basic premise of the movie is quite simple. Essentially it states that you can succeed, providing you have the talent and the tenacity to stick by your dream."[16] Although qualified by the admission that talent and tenacity are required, the thrust of this interpretation is that success is possible for everyone: the generic *you* can succeed. Similar sentiments are echoed in dozens of other reviews: "Anyone can cook. That is the motto of the great chef Gusteau. But what this really means is anyone can do anything"; "you can do anything no matter what size, no matter what, you can do anything"; "anyone can do anything they set their mind to"; "nothing is impossible, no matter who you are"; "if you work hard enough and never stop believing you can achieve anything."[17] Twenty-two users (10 percent) also mentioned artistic talent or accomplishment, but generally in a far more inclusive way than the professional critics. For example, one user wrote: "This movie shows that talent and art does not need a certificate, this can be found in anyone," and another said that the film "inspires you to be a great artist at whatever you do."[18] Only six IMDb reviewers (3 percent) echoed Anton Ego's interpretation of Gusteau's motto that not just anyone can be great, although greatness can come from anywhere. Another six actually revised Gusteau's motto to be even more inclusive than the original, claiming that he said that "everyone can cook."[19]

Clearly, in the film, not everyone can cook. Linguini is a disaster in the kitchen without Remy's guidance, and even Gusteau is responsible for recipes that are portrayed as disgusting without Remy's innovations. Interpreting the film as a parable about the universal potential to achieve anything is a stretch at best, and some people might even consider it a misreading. However, it's an understandable interpretation given the seductive power of meritocracy as one of America's master narratives and the cultural rewards of endorsing ideologies that legitimate the prevailing political and economic system. For those with a disproportionately large share of capital and power, meritocracies reinforce the belief that they deserve whatever they have. For those with a disproportionately small share, they reinforce the hope that if they work hard and act virtuously, they will eventually reap greater rewards. Even if meritocracy doesn't offer a guarantee of success for everyone, it may give people a sense of greater control over their lives. For everyone whose life and world view is shaped by the prevailing social system, meritocracies reinforce the idea that that system is basically fair and just.[20] This eagerness to say yes to meritocracy is especially conspicuous in audience responses to mass media texts that address perhaps the most powerful and tenacious meritocracy of the contemporary food revolution: the belief that eating the right foods in the right amounts and engaging in regular exercise will make anyone thin.

Selling the Meritocracy of Thinness

The Biggest Loser debuted in 2004, just months after U.S. surgeon general Richard H. Carmona declared obesity a national health crisis and "the

fastest-growing cause of disease and death in America."[21] The first season featured twelve contestants divided into two teams and promised a grand prize
of $250,000. The show was rushed to air to replace the foundering *Last Comic
Standing*, so it wasn't promoted with the rest of the fall lineup during NBC's
summer Olympics coverage. Nevertheless, 9.9 million viewers watched the
premiere. In an average minute of the broadcast, 10 percent of all adults aged
eighteen to forty-nine watching television were tuned in to *The Biggest Loser*.[22]
The third episode attracted 10 million viewers, nearly double the audience for
the premiere of Fox's heavily promoted reality show *The Rebel Billionaire*,
which aired at the same time. The fourth episode beat out the simultaneous
broadcast of CBS's Emmy Award–winning reality show *The Amazing Race*. At
its ratings height, the first season of *The Biggest Loser* performed better in its
time slot than any other non-Olympic programming on NBC that year.[23] As
of 2016, seventeen seasons have aired in the United States.[24] The size of the
audience fluctuates, ranging from highs around 17 million for some finales to
lows around 7 million for some midseason episodes, but it typically hovers
around 10 million.[25]

Most episodes follow the same basic formula. They begin with a recap of
the previous episode and contestants' reactions to the last elimination. Next,
the contestants are shown going about their lives in the mansion, where they
are recorded 24–7 in the style of reality television forerunners such as *Big
Brother* and *The Real World*. The first event in most episodes is the temptation, in which contestants have the opportunity to get a prize if they will eat
high-calorie foods. The temptation is usually followed by an exercise or diet
scene. In the exercise scenes, they're shown lifting weights and running on
treadmills, sweating and panting while their trainers bark at them to work
harder. In diet scenes, either the trainers or special guest chefs show the contestants how to prepare low-calorie, low-fat foods, often featuring sponsors
such as Jell-O and Jennie-O Turkey. Next is the challenge, in which contestants compete in some kind of physical contest. Prizes for temptations and
challenges fall into three categories: cash and other material goods, emotional
rewards such as a phone call home, or in-game advantages such as immunity
from elimination or a "pound pass" that can add or subtract from someone's
weight loss for the purposes of the competition. After the challenge, the contestants participate in the last-chance workout and then the weekly weigh-in.
Based on the results of the weigh-in, some of the contestants end up in the
elimination room, where they reveal the results of a secret vote that determines who will be sent home.

The very idea of a weight-loss competition presumes not only that weight
loss is desirable but also that individual weight-loss results can be compared

in a meaningful way. The popular belief that body size is a transparent reflection of eating right and exercising is part of the show's very foundation.
However, the show reveals many challenges to that ideology. The contestants
show little evidence of improper eating, even when presented with the "fattening" foods that are supposed to tempt them. Although eliminations are
supposedly based on weight loss, they are actually determined by secret votes,
which savvy competitors manipulate by forming alliances and using in-game
advantages strategically. Weekly weight-loss results are variable in ways that
cannot be entirely explained by diet and exercise, and even the show suggests that their weight loss is dependent on factors such as their gender, the
expertise of the trainers, the special equipment they have access to, and their
seclusion on the set of a reality television show with nothing to do but exercise. Ultimately, though, the show manages to fold all of these contradictions
into an overarching narrative of meritocracy, and that's the story that viewers
embrace.

Pathological Eaters and the Temptation Paradox

The Biggest Loser blames the contestants' weight on bad eating habits. They
are portrayed as weak-willed, emotional eaters who turn to fattening foods
for comfort even though they know there will be negative health consequences. For example, the title sequence for the first season begins with a shot
of a hamburger overlaid with the text "Do you have the will power?" The text
remains as the hamburger fades and is replaced by a fat male torso (see Fig. 12).
The image cuts back to the burger and then to another fat belly, this one with
a measuring tape wrapped around it, and then to a pile of doughnuts. The
images echo each other's curves, equating the round, squat, fattening foods
with round, fat bellies and bodies. The question implies a direct relationship
between having the will power to refuse foods like hamburgers and body size.
From there, the opening cuts to footage of all the contestants walking toward
the Biggest Loser ranch, implying that their presence there, like their fatness,
must be the result of consuming too many foods like hamburgers and doughnuts. The sequence reinforces the popular belief that people who get fat do so
because they eat bad foods in excess, unlike thin people, who are presumed
to eat them in moderation, if at all.[26]

The contestants are equated with bad, excessive food again when they are
introduced. After a brief preview of their future on the show, the host says,
"Let's take a look at your past" and signals for a wagon full of hay bales to pull
forward, revealing another set of hay bales covered with tablecloths and piles
of food. The contestants' names appear on placards that indicate which area

Figure 12. *The Biggest Loser* opening sequence. Source: *The Biggest Loser*, season 1, episode 1, aired October 19, 2004, on NBC.

of the pile corresponds to which person's past transgressions. Several of the contestants' mouths drop open, and some laugh uncomfortably. "The buffet is now open," the host quips. "This is some of the food that you all ate last week." The camera pans over piles of spaghetti and French fries. A contestant named Lizzeth[27] exclaims, "Oh my gosh" and covers her mouth in horror. They cut to an interview in which she says, "I ate all that? That is gross. It kind of made me sick, like, I can't believe I put that stuff in my body." In cuts away from a slow pan over buckets of popcorn, plates of doughnuts, and slabs of ribs, Kelly Mac says, "I didn't need to be eating all that food. I'm five feet tall!" Maurice describes the food as being "laid out like a body being laid out such as a wake or at a funeral," and says that saying goodbye to it, "was like saying goodbye to a loved one." The host says, "After stepping through those doors you will no longer have this to comfort you."[28] However, none of the contestants seems to regard the food as comforting. Even Maurice, who compares the food to a loved one, says it's like a dead body. Instead of expressing a desire to eat the foods that have supposedly made them fat, they respond with the same shock and disgust the display seems designed to inspire in the television audience.

It turns out the host's promise that they'll have to bid the bad foods farewell is only partially true. The foods in the buffet reappear on the show in almost every episode, often as representations of the contestants. For example, the elimination room is lined with glass-front display refrigerators with the contestants' names at the top. When they first encounter the refrigerators, the host says, "Inside each of your refrigerators is your biggest enemy: temptation." The camera zooms in on a basket of fried chicken inside the "Maurice" refrigerator and then cuts to an interview with Maurice in which he says, "I'm not even gonna lie to you. Fried chicken: that's a southern boy's favorite."[29] The refrigerators remain illuminated as long as the contestant remains on the show, and if they are eliminated, the light dims.

Late in the season, the show sometimes shifts from equating bad foods with the contestants to equating them with the weight the contestants have lost. In season 1, episode 8, the five remaining contestants are confronted with life-size cardboard cutouts of themselves as they looked when they arrived at the ranch. Each one stands next to a pile of their favorite food equivalent to the amount of weight they have lost so far in the competition: 77 pounds of apple pie, 48 pounds of pizza, 50 pounds of fried chicken, 45 pounds of macaroni and cheese, and 44 pounds of spaghetti. The platforms are too small to hold it all, so some of it spills onto the ground, emphasizing how excessive and disgusting all this food is meant to look. The host says, "You guys, you've all been carrying that weight around for so many years. And it feels so much better to be that much lighter."[30] Unsurprisingly, they all look appalled rather than enticed.

Despite their supposedly pathological appetites and their emotional attachment to fattening foods, the contestants virtually never express any desire to eat them, even when given the opportunity. On rare occasions, usually early in the season, a contestant will have a teary meltdown because he or she feels hungry or is confused about what he or she is supposed to eat. In season 1, episode 1, a contestant named Lisa is shown sitting at a table in the kitchen and crying as she says, "I'm just hungry. I don't know what to eat. I think I just overate chicken. I don't know how many ounces it was and then I put barbecue sauce on it. . . . I ate 595 calories in one day. That can't be enough." Later in the same episode, the contestants awaken to a buffet of breakfast foods that includes French toast, butter, syrup, and turkey bacon. Maurice is shown repeatedly reaching for the bacon, and his teammates shun him for the rest of the day.[31] But those scenes are rare exceptions. While they are on the show, women are counseled to eat between 1,200 and 1,500 calories per day and the men between 1,500 and 1,800 calories per day and for the most part, the contestants are portrayed as obedient adherents to their restricted diets.

Indeed, they often express surprise at how delicious or satisfying they find low-calorie, low-fat alternatives. In season 3, trainer Kim shows her team how to prepare a sugar-free dessert she calls Jell-O Chocolate Berry Bliss, and several of the contestants say it's something they can see themselves serving to their friends, who won't even know how many calories they're saving.[32] In other episodes, they grill turkey burgers, make pizzas on whole wheat tortillas, and share a Thanksgiving meal made virtuous with low-fat Jennie-O Turkey and artificially sweetened pumpkin pie, all of which are framed as proof that eating healthfully can still be satisfying and delicious.

However, that doesn't prevent the show from portraying the contestants as pathological, emotional eaters. In the third season, trainer Bob enters the kitchen where his team is preparing dinner after having lost that week's challenge. He asks them what they're doing and they tell him they're preparing some white fish, which is typically seen as a good food. The show cuts to an interview with Bob, who says, "You talk to overweight people and a lot of them are emotional eaters. And when you're dealing with emotional eaters, when anything bad happens, they're gonna reach right for the food. Especially when you're dealing with people that are the size of the ones that are in the house right now." Apparently because of their size, the contestants must be eating emotionally, even if they're making what most people would see as a virtuous meal. Bob tells them that instead of eating dinner, he wants them to get in another workout. One of the contestants asks, "Before we eat?" Bob nods. "Before you eat." Although they seem stunned for a moment, perhaps contemplating the cold fish they'll be returning to afterward, they dutifully follow him to the gym.[33]

The contradiction between how contestants actually react to bad foods and the show's insistence that they're emotional, out-of-control eaters reaches its apex in the temptations. In almost every episode, contestants are presented with the opportunity to eat fattening foods, but they almost never express a desire to do so. Instead, they agonize over whether the prizes are desirable enough to justify the extra calories. Emotional rewards, such as the opportunity to speak to their distant family members, cause the most consternation. Contestants also carefully weigh in-game advantages against the estimated caloric content of the food they would be able to eat, sometimes explicitly calculating how many minutes of exercise they would have to do to make up for taking the bait. Instead of seeking solace in food, the contestants express emotional attachments to their loved ones. They engage in what appear to be coolly rational calculations about how the foods will affect their weight-loss goals. The food is actually a disincentive. The contestants may want to eat it, but even the way the feature is structured acknowledges that actually getting them to do so requires a motivation beyond any inherent desire they might have to eat the foods they're supposed to be so bad at resisting.

Despite the additional incentives, contestants almost always decline the temptation. Even in the first episodes of every season, when the contestants are portrayed as self-control neophytes who still need to build up their willpower along with their muscles, they usually resist. In season 2, episode 2, the contestants have to sit at a dinner table where letters from home peek out from underneath platters full of fattening foods. They are told they can only read the letters if they eat what's on the plates. The camera lingers on contestants who lick their lips or stare at the food intently. However, when the contestants talk about the food, they say they don't want to eat it and express frustration that they would have to sabotage their diets in order to read the letters. Shannon is so distraught over not being able to read the letter that the other contestants begin encouraging her to eat the food, telling her she can work it off in the gym later. Ultimately, even Shannon resists.[34] In other temptations, when at least a few of the contestants decide the prize is too good to pass up, as is often the case when they have the chance to win immunity from elimination, contestants eat the food quickly and without much apparent pleasure. They also confer with each other to estimate the calories they're consuming and come up with strategies for how to make up for it later.

Nevertheless, the host consistently frames temptations as a test of contestants' ability to resist their unruly desires, which are portrayed as an "enemy" they all must conquer in order to become thin. Shots of contestants licking their lips are often featured in recaps, frequently taken out of context to suggest that it really is the food that's tempting, not the additional prizes, despite what the contestants say and do. The show also routinely frames the temptations as preparation for the "real world" and all the desirable foods they will encounter there, obscuring the role of the prizes in getting them to eat the bad foods on the rare occasions when they do. Furthermore, when contestants do decide the prize is worth taking the caloric hit, capitulating to the temptation is portrayed as evidence of deep-seated psychological issues with food, not a rational calculation aimed at winning the competition.

For example, in season 4, episode 3, the host introduces the temptation as a "real-world challenge." She tells the contestants that the show has filled a room with their favorite foods, like pizza, brownies, and cupcakes. Each of them will spend four minutes in the room, and whoever eats the greatest number of calories will win a three-pound pass for their team. The host also says that no one else on their team will know how much they ate. Only two of the contestants eat any food while in the room, and both do so in a deliberate fashion that belies the notion of succumbing to temptation. Once in the room, Neil begins grabbing at whatever is closest to him without pausing to select favorite foods or seeming to enjoy them; at one point, he tips a small bowl of M&Ms into his mouth like a shot of liquor, a moment that is replayed several times in the episode recaps. Patty also eats as if on a mission once she gets in the room, indiscriminately shoving egg rolls, eclairs, and handfuls of apple pie into her mouth. Both later claim that their primary motivation for eating the food was to win the three-pound pass for their team.[35]

In contrast, consider the behaviors demonstrated in the classic studies of willpower and delayed gratification known as the Stanford marshmallow experiment. The researchers, led by psychologist Walter Mischel, left children ranging from three to five years old alone in a room with a small desirable reward such as a marshmallow placed on the table in front of them. The children were told that if they waited to eat the treat, they would get a second one in fifteen minutes. While waiting, many covered their eyes with their hands or physically turned away from the table. Some started kicking the desk, fidgeting with their hair, or singing songs. Many examined, smelled, or stroked the marshmallow covetously, sometimes just before succumbing to temptation. Only one-third of the children delayed long enough to get the second treat, but most of the two-thirds who succumbed to temptation waited at least half a minute. Those who were offered a "cognitive distractor" such as a toy or instructions to "think about fun things" waited an average of eight and a half and twelve minutes, respectively.[36] In the temptation, the contestants who didn't eat any of the food display many of the same behaviors as the children in the marshmallow experiment. Some squeeze their eyes shut, physically turn away from the food, or roll their eyes toward the ceiling in an apparent attempt to avoid looking at it. A few eventually lean toward the table, examining, smelling, and sometimes even touching the food. That's what willpower being tested looks like. The two contestants who do eat engage in no gratification-delaying behaviors and show no signs of capitulating to a desire that runs contrary to their rational intentions. Their consumption instead appears deliberate and premeditated, an act of will instead of a failure of it.

Both the symbolic use of food to represent the contestants' bodies and the temptations are ways of getting bad foods onto the show without contestants having to eat them very often. The way those foods are used in the show promotes the idea that fat people look the way they do because of their excessive, inappropriate appetites. Although contestants routinely turn down the opportunity to eat the bad foods or do so only when compelled by nonfood incentives, the few occasions when they do eat the food and any footage that seems to represent their desire for the food is played repeatedly, divorced from its original context, to reinforce the belief that fat people have irrepressible appetites. All of this works to further the ideology that body size is primarily dependent on self-control and thus that fat people are fat primarily because they lack the willpower to eat the right foods in the right amounts.

Making the Competition Seem Fair

The Biggest Loser portrays eliminations as merit based even when they are based on a group vote or participation in a single workout, as they were in the first episode of season 3. For that season, NBC recruited fifty contestants, one representing every state. All fifty of them appear in the premiere, and in the opening minutes of the episode, the host tells them that only fourteen will be chosen to stay at the ranch. Then she introduces the personal trainers, who immediately begin leading the group in a calisthenics routine. In between shouting instructions to the crowd, the trainers run up and down the rows of furiously jumping and squatting contestants, asking them questions such as "Do you want this real bad? How bad?" After the workout, the trainers return to the stage and choose seven contestants each, alternating like team captains in grade school dodgeball.

Although the trainers' choices were probably determined in advance by the show's producers, the show frames getting to stay at the ranch as a reward for hard work. Each time the trainers call one of the contestants to the stage, they claim to have seen something special in them during that single work-out, which they typically call "heart" as the show cuts to a clip of the chosen contestant exercising and looking pained. Interviews with the contestants also seem to endorse the idea that the workout is a meaningful test. The woman representing Vermont, who was not chosen, says, "I thought I was going to *die* when we did the first workout in our rows. I kept thinking, like, you gotta do it, just one more time 'cause they're watching you and you know, maybe if I just get my legs up a little higher, I'll be one of the fourteen to stay." However, most viewers who wrote about the episode in online forums were not convinced, many complaining about how arbitrary the mass elimination seemed.

Individual eliminations are also portrayed as merit based, usually with somewhat more success. Later in the same episode, the red team loses the first weigh-in, meaning they have to vote to eliminate one of their team members. Afterward, the team is shown back at the mansion, reacting to the news. Contestants Jen and Heather have both lost four pounds, but Jen's start-ing weight was lower. Heather points at her accusingly: "You had the lowest percentage," she says, "So just know that. So don't try and lie to yourself or lie to anyone else. You had the lowest percentage because you didn't bust it." The show cuts to an individual interview with the team's trainer, Kim, who says that at first she was "shocked" that Heather "called Jen out," but that on further reflection, "I could kind of see it. I told Jen several times this week, 'Step it up.'" Then a flashback plays showing the red team at the gym earlier in the week. Jen appears to be struggling on an elliptical machine, and Kim comes over and asks, "What are you doing?" Jen says wearily, "I think it's called a break," and Kim barks, "Uh, there are no breaks here." The scene cuts back to the house, where Kim looks intently at Jen as she says, "Only you know if you're working 100 percent." But the implication is clear: the scale has revealed the truth, and Jen wasn't working hard enough. At the end of the show, her team members vote to send her home.[37]

However, the contestants' bodies don't always cooperate with the creation of tidy narratives about weight loss as a reliable measure of willpower and hard work. Some weeks, contestants don't lose any weight or even gain weight, despite spending hours in the gym and apparently eating mostly fat-free turkey and sugar-free Jell-O. In season 1, episode 2, all of the competitors lose much less than they did in the first week and several lose nothing at all or gain weight back, causing trainer Jillian to hypothesize that they are eating so little that their bodies are hoarding calories and they will need to eat more to trick their bodies into losing weight again.[38] When a contestant named Wylie loses nothing in season 3, episode 9, trainer Kim tells him it must be because he's building muscle so fast, which weighs more than fat.[39] In general, contestants alternate between good weeks, when they lose a lot, and bad weeks, when they lose a little. Some contestants deliberately try to have a bad week when they're protected from elimination by immunity in order to set themselves up for a good week when they might be vulnerable again. For example, in season 3, Erik gains three pounds while he has immunity in episode 7 and then drops twenty-two pounds the next week.[40] Similarly, in season 4, Neil gains seventeen pounds while he has immunity in episode 7 and later admits that he did it by drinking a lot of water before the weigh-in.[41]

Nevertheless, faith in the energy balance paradigm is so strong that fans of the show generally attribute almost all the contestants' weight changes to the inexorable logic of calories in and calories expended. For example, in a post on the online community Fat Fighters, a contributor offered to "crunch the numbers" to figure out how dramatic losses, such as Amy's sixteen-pound loss in season 3, episode 1, would be possible: "At 260 lbs, she needs 2333 calories to maintain her weight if she is sedentary. Bob put his ladies on a 1200–1500 calorie diet, creating a 1100 daily deficit via eats alone. . . . I am going out on a limb and say they workout intensely 4 hours a day. . . . @ her size, she can burn 5000 calories in those 4 hours if she just walked or jogged. . . . Easily she burned 6000 calories. 6000 burned via exercise + 2100 daily deficit ='s doing the above for 7 days @ her 260 lb weight is how she shed 16 lbs in one week."[42] Irregularities such as good weeks and bad weeks, muscle weighing more than fat, and how much water the contestants might have consumed rarely figure into attempts such as these to make sense of weight loss through calorie math.

Differences in weight loss based on gender have prompted the most active debate about the fairness of the competition both in the show and in its online fan communities. Men typically start off heavier and lose weight faster. When the first two members eliminated from the blue team in season 3 were women, Bob told the one remaining woman on his team that she would have to work out "like a man" to stay in the game.[43] In the same season, a red team contestant named Wylie expressed jealous admiration for his team member Kai, who "loses weight like a guy."[44] Kai was the runner-up for the grand prize, and at the finale the host praised her for losing the most weight "of any woman in the history of the *Biggest Loser*," implicitly recognizing that the bar was lower for female competitors.[45] In all of these instances, the show suggests that the playing field is not level for men and women.

Unsurprisingly, the show's occasional attempts to pit women against men have largely failed. In season 2, the contestants were initially divided into teams based on gender, but the teams had to be rearranged after the women's team lost three weigh-ins in a row. The battle-of-the-sexes theme remained part of the season's continuing narrative, and special attention was devoted to the fact that two of the final four contestants were women, which was supposed to prove that the competition does not inherently favor men. However, viewers generally seem to believe that underlying biological differences predispose men to succeed in the competition, and they are divided about how that affects the show's fairness. In December 2007, after a man won the grand prize for the fourth consecutive season of the show, a member of NBC's online *Biggest Loser* community started a thread titled: "IT'S TIME FOR A FEMALE TO WIN!" The original post elaborates: "Julie was the first woman who had a good opportunity to win and she still didn't. She looked fantastic and could not have lost any more! I'm frustrated and it's time for the show to do an all-female cast." Another forum member replied, "I'm beginning to realize that it's going to have to be an all women final 3 or final 4 for a woman to win."[46] Most people who responded to the thread agreed that women have an inherent disadvantage that makes the competition unfair if they have to compete with men.

However, a sizable contingent disagreed. They pointed to the fact that at least one woman has made it to the finals in every season, and many of them claimed that a victory based on different rules for women would be unsatisfying. One person recalled, "Last year, Poppi beat out 35 other people to win [the "at home" prize] by losing over 50% of her starting weight—if she had been in this year's final four, she would have been crowned the Biggest Loser with that stat!" Another said, "I want to see a female biggest loser to, but I want to see her win it fair and square. If they changed to rules to help the women, I wouldn't feel good about the win. It can be done . . . and I'm sure it will." Some argued that implying that gender is an issue undermines the achievements of the winners: "Your frustrated that a woman has not won? I am frustrated that you are trying to cheapen the wins of every winner by claiming that the competition was not fair. They won fair and square and they had amazing accomplishments."[47] In season 6, Ali Vincent became the first woman to win, silencing most complaints about any inherent gender inequality in the competition, and after seventeen seasons, six women have won the grand prize and another six have been crowned winners of the "at home" competition.

Debates about whether weight loss is based on merit alone largely eclipse debates about whether the competition is structured to reward weight loss in a fair and reliable way. In most seasons, contestants initially compete in pairs or teams. Anyone on a losing team can be eliminated, even someone who has lost impressive amounts of weight. In season 1, Aaron accused his teammates of voting him off because he was losing weight so quickly he was a threat to win the competition.[48] After the teams are dissolved and the remaining contestants compete individually, they sometimes form alliances with each other or favor their former teammates in the elimination room. Attempts to manipulate the elimination votes are often referred to as "playing the game" and are generally shunned, especially by the trainers. However, contestants also claim that you have to play the game if you want to win.

Viewers who participate in online forums also tend to criticize "game playing" and often express frustration when the show doesn't reward maximum weight loss. In most seasons, only the contestants who remain on the show until the end are eligible to compete for the grand prize, although the eliminated contestants and "at home" contestants are eligible for lesser prizes. The result is that the "biggest loser" is not always the person who has lost the most weight by the finale. The first time an eliminated contestant lost a greater percentage of his starting weight than the season winner was in season 4. Many participants in the NBC forum claimed that the result was dissatisfying or unfair, some declaring the entire season a fraud. For example, one wrote, "On all the other seasons the winner of the show could say with conviction that they lost the most weight. Not this season. The winner of this season can only say he won it but did it through game playing."[49] In several subsequent seasons, one or more eliminated participants again lost more weight than the champion by the finale. However, the show never chose to highlight those instances, so it was likely only apparent to fans who were calculating the final weight loss percentages themselves. In general, the show tries to portray the results of the competition as an accurate reflection of contestants' determination and hard work with the scale as an impartial judge, and audience members generally seem to play along, ignoring any inconsistencies in how people's bodies respond or how the show rewards them.

Both the show and its viewers sometimes struggle to reconcile the popular belief that anyone can lose weight through diet and exercise with the fact that the contestants have exceptional resources and incentives. Indeed, many of the contestants identify the show as a unique opportunity or say it is their last chance, a potentially humiliating and painful trial they are willing to go through only because all of their previous weight-loss attempts have failed. They often claim that they could not succeed on their own. The trainers also sometimes acknowledge that the weight loss people achieve on the show may be impossible for anyone to replicate at home and difficult for the contestants to maintain once they return to their normal lives.

In season 1, episode 10, the challenge involved holding on to a bunch of helium-filled balloons. The prize for the last person to let them go was a professional-grade treadmill just like the ones in the *Biggest Loser* gym. In individual interviews, the contestants discussed the significance of the prize. Kelly said, "If there was one piece of equipment I would love to take home it would be the treadmill," and Ryan emphasized the dollar value of a professional piece of fitness equipment: "The fact that the treadmill was the prize was definitely motivation enough to win this. Jillian tells us they cost seven or eight thousand dollars."[50] All of the contestants expressed concern that they would not be able to maintain their weight loss without access to the kind of equipment that was available to them on the show. The show frequently features scenes that specifically address that anxiety. At least once a season, the contestants and trainers arrive at the gym to find it padlocked. They exercise outside to prove that you don't have to go to the gym to get a great workout. Special field trips are also framed as practice for the "real world." In season 3, the trainers took their teams to a beach, where they did yoga and ran through the sand, repeatedly framing the experience as proof you can exercise anywhere.[51] Later that season, the remaining contestants went on a cruise where they were tested on their ability to eat moderately at the ship's buffet and work out with the resources available on the ship.[52]

Some viewers, such as Sarah Dussault, who writes columns on *Diet.com* under the moniker "the Diet Diva," criticize the show's failure to offer a more realistic role model for real-world dieters. In a column titled "Biggest Loser: Do They Eat?" she wrote, "Working out 8 hours a day is not something you or I can do. Why do they show us footage of things we can't take away any tips from?" A comment on the column agreed: "Just started watching it this season, and while I love it and feel that it's really inspirational for a lot of people, I don't know that it's setting the best example in some ways. Pretty much no one can lose 10 lbs a week in 'the real world' (which I suppose they do mention)."[53] Members of the *Diet.com* message board expressed similar concerns: "I think they lose so much b/c of the 24/7 attention to diet and exercise. We can all diet but not many have the luxury of several intense workouts in one day! Some of us have to work!"[54] However, both the show and many viewers hold up contestants who achieve significant weight loss at home after being eliminated as evidence that people can achieve dramatic weight loss even without special resources. Later in the *Diet.com* thread, another user referred to one of the contestants who was eliminated in season 3: "I[t] goes to show that if [Ken] can lose that much while at home, they aren't giving them some magic potion. When people ask me what I am doing to lose weight, I say, diet and exercise. So simple that it works. I know I can, I know I can."[55] Many participants in this and other forums echoed that sentiment.

The show also offers a conflicted message about the extent to which the personal trainers deserve credit for the contestants' weight loss. Sometimes the personal trainers are portrayed as essential. For example, one of the penalties for losing a challenge or refusing a temptation is being denied access to the personal trainers, which is portrayed as a serious handicap. In season 3, when Erik lost access to a personal trainer for forty-eight hours, his team's trainer looked at him and said, "You're screwed."[56] However, Erik had a good week at the weigh-in, which was framed as evidence that he could lose weight on his own. Later in the season, the "no trainer" episode was portrayed as a turning point for Erik, the moment when he took control of his life and his weight loss.[57]

When two of the "at home" competitors rejoined the remaining competitors at the ranch halfway through season 3, they struggled with the trainers and challenged their expertise and authority. Both returning competitors lost a greater percentage of their starting weight at home than anyone still at the ranch, and they openly questioned whether following the trainers' advice would impede their progress. Trainer Bob expressed concern about Jaron's food diary, claiming that his diet wasn't sustainable, but Jaron protested that it was working for him. After a few frustrating experiences working out with the red team and trainer Kim, Adrian chose to exercise on her own. In interviews, she claimed that Kim didn't understand her body and that she was afraid that if she followed her advice, she would get eliminated.

The show's ambivalence about whether the trainers are essential or peripheral came to a head in the emotional scenes where the finalists said good-bye to them before returning home for the final sixty days of the competition. In season 3, the eventual champion Erik admitted to trainer Bob, "The one thing that makes me sad about leaving here is you." Bob reassured him, saying, "You are a man in control of his life again," and introduced Erik to a cardboard cut-out of himself as he looked when he first appeared at the ranch. Erik began to cry. He told Harper, "You said you were going to save my life and you did." Bob encouraged Erik to talk about how he felt when he was that big and a morose soundtrack played while Erik talked about how unhappy he was. Then Bob turned the conversation toward Erik's accomplishments and the soundtrack began to soar hopefully. He announced that Erik had lost 124 pounds, more than any previous contestant in their time on the ranch, and said, "You did it. My friend, you did it all. And that's what I am so proud of. Look how far you have come. Because of your determination and your focus." Erik cried, nodded, and accepted the praise and responsibility for his weight loss.[58] Although the show portrays being on the show and getting to work with professional trainers as helpful tools, in order to support the ideology that body size is primarily determined by an individual's determination and hard work, it must ultimately shift responsibility from those tools onto the contestants' shoulders.

TUNING IN FOR SUCCESS, NOT SCHADENFREUDE

The weigh-in is the longest segment of *The Biggest Loser*; it constitutes the second half of each regular episode and the entirety of the season finales, which are two-hour live broadcasts. Contestants don special outfits and step onto a giant scale with large flat-screen displays behind them and a smaller display mounted at their feet. The outfits put their bodies on conspicuous display: all of the contestants wear shorts, and women wear a sports bra while men wear T-shirts that they remove before getting weighed. Literary scholar Jennifer Fremlin compares the moment they step on the scale to the money shot in porn, a normally private moment of vulnerability turned into a titillating spectacle. Fremlin says she fast-forwards through the rest of the show to get to this segment.[59] Based on the Nielsen ratings for season finales, she's probably not the only one. Ratings grow substantially in the second hour, when recaps of the season finally give way to people getting on the scale.[60] According to Fremlin, weigh-ins offer viewers a kind of squirming, illicit pleasure: "They put their shame on display: at being fat, their exerted bodies wheeze and squeeze into spandex workout clothes meant for svelter shapes. Their exhibition becomes a cover for our own shame as viewers who, by participating in their humiliation, in turn abject ourselves."[61] In her reading, everyone is ashamed—the contestants about their fatness and the viewers about taking pleasure in watching people be humiliated because of their fatness.

Many viewers probably do see the weigh-in outfits as humiliating and find reassurance in the fact that they're thinner than the people on the show,[62] but the weigh-ins are also designed to encourage viewers to identify with the contestants. Like the contestants, the viewer waits in suspense as a series of false numbers flashes on the displays. Much of the drama and appeal of the show is based on the contestants achieving big weekly losses, so audience members are also invested in seeing the numbers go down. When the scale finally settles on a number and the net loss and percentage loss are displayed, contestants often pump their arms triumphantly and whoop with joy. Perhaps some people see them as pathetic or grotesque even at that moment, but many others are celebrating with them.

Midway through each season, the weigh-in outfits change. Men wear sleeveless T-shirts that hide any remaining belly fat and the women wear form-fitting tank tops that streamline their bodies. Beginning with those episodes, the weigh-in is shown with a split screen. One side shows a video of them from the first weigh-in, fat and exposed, next to their more forgivingly attired bodies. The new outfits and split screens emphasize the visual evidence of the contestants' hard work and success. Echoing the conventions of "before" and "after" weight-loss advertisements, the earlier weigh-in images often show them looking dejected or chagrined, providing a contrast with the usually happy response to their new weight (see Fig. 13). People who fast forward or turn the show on just to see the weigh-in are tuning in for that, too.

Fremlin's reading of the weigh-in reflects a broader consensus among critics of reality television that the format succeeds primarily by putting people in uncomfortable situations in order to draw viewers who are attracted by the ignoble act of watching other people suffer. The genre is often accused of reinforcing undesirable social behaviors. Media studies scholar Susan Douglas sums up critiques of the genre: "We should appreciate that reality TV, particularly, traffics in and relies upon voyeurism, one-upmanship, humiliation, and often soft-core pornography. . . . It exhorts us to be a voyeur of others' humiliation and to see their degradation as harmless, even character-building fun."[63] Some scholars argue that viewers derive satisfaction from seeing people fail because they get to feel superior to the losers. For example, a 2004 survey at Ohio State University found that respondents who reported that they watched and enjoyed reality television had "above average trait motivation to feel self-important and, to a lesser extent, vindicated, friendly, free of morality, secure, and romantic," which the authors interpreted as support for the theory that viewers watch reality TV primarily to feel better about their own lot.[64]

Cultural critic Lee Siegel offers an alternative theory, but one that nonetheless depends on the assumption that reality television primarily portrays people as failures. He claims that "only in America" could reality television become a "gigantically profitable object of diversion" because most Americans feel like losers and thus identify with others who fail: "[Reality] television consoles people for their daily failures and defeats rather than making them feel superior to other people's failures and defeats. Reality television replaces

Figure 13. Mid-season weigh-in. Source: *The Biggest Loser*, season 13, episode 11, aired March 13, 2012, on NBC.

the glowing, successful celebrity ideal with gross imperfection and incontrovertible unhappiness. In a ruthlessly competitive society, where the market has become the exclusive arena of success, reality television shames the illusion of meritocracy by making universal the experience of the underdog, the bumbler, the unlucky and unattractive person."[65] This theory doesn't explain the appeal of *The Biggest Loser*, or of most other makeover and competitive reality shows for that matter. This subgenre of reality television emphasizes the transformations people achieve by submitting to expert advice. Although only one contestant every season wins the big prize and eliminated contestants are often accused of not working hard enough, the losers aren't portrayed as abject. Many continue to lose weight for at least a while after leaving the show, and in the updates shown after the elimination, they often talk about how they have improved their diet and exercise habits. For example, first season contestant Maurice lost an additional twenty pounds between the time of his elimination from the ranch and the broadcast of the episode when it happened. At 363 pounds he may not have looked significantly slimmer than he did at 380, but in his update, he said that he was continuing to eat "lighter" and exercising to meet his weight-loss goals. He also shared the news that he had been appointed to serve as a spokesperson for a Tennessee Department of Health program called "Respect Your Health."[66] Both the hosts and the contestants frequently riff on the series title, noting that by losing weight, whether they win the grand prize or not, they are really all winners.

Instead of "shaming the illusion of meritocracy," the way Siegel claims reality TV shows do, *The Biggest Loser* affirms viewers' faith in meritocracy. Dozens of critics, bloggers, and message board participants claim that the show has inspired them and convinced them that they too can lose weight even if their past attempts have failed. Some people even report being moved to exercise while watching the show. A post on the blog *Brand Liberators* by Brenda Rizzo exemplifies the dominant audience response: "I love *Biggest Loser*. I watch it faithfully every season. My husband wonders why. Is it because of the game playing, or tips on eating or the tips on exercise? . . . And what about those outfits? Why do they make the men take their shirts off to get weighed when they are at their heaviest? . . . The real reason I love *Biggest Loser* is for the sheer fact that it works. I have struggled with weight my whole life and I just love to see someone have success with weight loss. For the most part—the past contestants have kept it off. Who else has statistics like that?"[67] Instead of claiming to take pleasure in the unflattering initial weigh-in outfits, Rizzo criticized them. She said that her struggles made her appreciate seeing someone succeed, and although she seemed to be aware that most weight-loss diets fail, she embraced *The Biggest Loser* as proof that success is possible. Rizzo didn't claim to have lost a significant amount of weight herself, but she clearly wanted to identify with the winners.

Perhaps the best evidence that *The Biggest Loser* is seen more as inspirational than humiliating is the success of the brand, which grossed an estimated $100 million in 2009.[68] Many of the books in the Biggest Loser Cookbook series have spent weeks on the *New York Times* bestseller lists.[69] There are *Biggest Loser*–branded bathroom scales, kitchen scales, workout DVDs, stability balls, resistance bands, clothes, and a line of protein powders, shakes, and bars.[70] There's an official online weight-loss club that charges approximately $20 per month for access to a variety of interactive tools. It advertises that "you don't have to be on the show to lose weight and change your life!"[71] Finally, for people willing to pay to get the same treatment as the contestants minus the cameras, there are three Biggest Loser resorts, in Utah, California, and New York. Visits start at $1,700 per week for off-site residency or $2,000 to stay at the resort.[72] People are literally buying into *The Biggest Loser* and the promise of success it offers.

Despite the conviction fans such as Rizzo express that *The Biggest Loser* really works for long-term weight loss, that is not entirely supported by what little information is available about what happens to the contestants after the finale, which is filmed right around the typical weight-loss dieters' nadir, six months after the contestants first arrive at the ranch.[73] In every season, at least one contestant has lost over 100 pounds. In season 3, nine of the contestants from the ranch and another nine of the contestants competing from home lost that much, and the winner lost over 200 pounds. Although these results may not be typical for weight-loss dieters, what follows for most of them is that they begin to regain the weight they lost. The executive producer and physical trainers told the *New York Times* in 2009 that half of the contestants remain close to the weight they achieved on the show for several years. A study that followed fourteen Biggest Loser contestants for six years after the end of the competition found that all but one regained a significant amount of the weight lost during the competition. Five of the study participants returned to within 1 percent of their starting weight or above it. Despite substantial weight regain, their average resting metabolic rate remained suppressed, burning an average of 500 kcal per day less than expected for their weight and age.[74]

Contestants sign contracts before participating in the series that prevent many of them from talking to reporters. Nevertheless, several contestants have risked fines and disgrace by publicly admitting to regaining almost as much weight as they lost on the show. Ryan, the first champion, told the *New York Times* that he had not been invited to the special reunion episode in 2009 and speculated that he had "been shunned by the show because he publicly admitted that he dropped some of the weight by fasting and dehydrating himself to the point that he was urinating blood."[75] The following year, he was invited to participate in a reunion episode as the one example of a failed former contestant. He admitted to being just ten pounds shy of his starting weight when he went on the show. His former trainer Jillian hugged him but then called him a jerk and said he had learned nothing from her. Ryan took personal responsibility for his weight regain, saying that he struggled to incorporate the lessons from the show into his life at home but intended to do better and lose the weight again.[76]

One contestant has made good on a similar promise. After season 3 winner Erik regained most of the 200 pounds he lost on the show within two years of the finale, Discovery Health produced an hour-long documentary about him titled *Confessions of a Reality Show Loser*.[77] In the documentary, his trainer from the show showed up at his house unannounced and challenged him to return to *The Biggest Loser* to weigh in during the season 9 finale. Erik accepted the challenge, and at the finale, Allison announced that after spending five weeks at the Biggest Loser resort, Erik had lost 150 pounds.[78] Again.

Another contestant who has publicly admitted to regaining weight after the finale but who is less contrite about it is season 3 runner-up Kai. She began writing about the dehydration techniques used to produce the big numbers on the scale on a MySpace page in 2007. She claimed that while on the show, she exercised in multiple layers of clothing and drank no water for twenty-four hours before weigh-ins. Before the finale, she binged on asparagus for its diuretic effect and had a colonic.[79] In 2010, she appeared on *The Early Show* on CBS, claiming that she had regained 30 of the 118 pounds she had lost on the show in less than a month after the finale merely by drinking water normally again. By the time of the interview, she had regained a total of 70 pounds and said she was struggling with an eating disorder as a result of her experiences on the show.[80] In an article about Kai's accusations in the *St. Petersburg Times*, her team's trainer was quoted as saying that "she once saw a black trash bag sticking out from under one competitor's sweatshirt during a work out" and that the producers "love big numbers on the scale" and downplay the role of dehydration in achieving them.[81]

Even *The Biggest Loser* reunions reveal that many contestants regain weight after the finale. However, those episodes obscure the trend somewhat by focusing on contestants from recent seasons, who may not have gained as much weight back yet, and by comparing contestants' current weight to their starting weight when they entered the competition instead of their weight at the finale. Each reunion episode also features at least one contestant who has regained most or all of their initial weight loss, but it portrays them as the exception, not the norm. Thus, despite the producers' concession that only about half of the contestants are able to maintain most of their weight loss, the reunion shows paint a picture of far more consistent long-term maintenance.[82]

Although the reunion shows do their best to provide an optimistic view of long-term weight-loss maintenance, for many viewers even the supposedly successful former contestants do not live up to the show's promise of universally achievable thinness. A recap of one reunion episode on the website *Television Without Pity* is peppered with comments such as, "After a commercial break, we learn that five former contestants have decided to get off of their couches and get to work battling the childhood obesity epidemic. . . . Ed should maybe work on treating his own self again, because he's looking a touch hefty," and "Sione is clearly very moved by his experience in Tonga. . . . At this point it feels tacky to mention that he's maybe put on a couple of pounds." This writer also noted: "Is nobody going to remark on the fact that the winners of seasons one through three have all plumped back up? This seems like a problematic trend." Nonetheless, the *Television Without Pity* recap ended by claiming that the episode was inspiring and described the former contestants as "looking fly."[83] At least for the recap author, the overall narrative of triumph overwhelmed the seeds of doubt planted by the signs that many contestants had not maintained their weight loss.

The message boards on *Television Without Pity* reveal a wide range of reactions to reunion episodes. Some people argued that on the whole, the results are obviously good or "speak for themselves," while others suggested that some regain is natural but express concern about former contestants who seem to be "slipping." After the second reunion show, one user noted that only one of the contestants had kept off all the weight lost on the show so far and that two contestants who had become personal trainers weighed thirty-five to forty pounds more than at their finales less than a year earlier. "If they couldn't manage to keep their own weight off, how are they going to help other people reach their optimum weight?" they asked. Others responded that some regain is inevitable and maybe even healthy, but the original author protested, "Most contestants are at a healthy weight at the finale, so if they regain 30+ lbs, they're coming close to the overweight territory again. I guess I was just hoping that, once the media circus is over, the contestants would continue working towards/staying at a healthy weight for their own sake. I don't mean keeping up the 8hr/day workouts, but a consistent regimen of intense workouts 3–4 times a week, coupled with a calorie-controlled diet, should get 99% of the population to a healthy weight eventually."[84] The viewer expressed absolute faith in the meritocracy of thinness. Even though the contestants in question have become personal trainers and thus likely exercise regularly as a part of their jobs, he or she assumed that they must be doing something wrong if they're gaining weight. Although this criticism of the contestants reads against the grain of the reunion show, which holds former contestants up as success stories, it's precisely the unforgiving attitude toward body size that most *Biggest Loser* audiences embrace.

Choosing Meritocracy in Spite of the Evidence

Given that *The Biggest Loser* brand depends on people believing that its methods produce lasting weight loss, the misleading portrayal of long-term weight-loss success might be seen as a craven form of mass media manipulation. Critics of the public health campaign against obesity often point to the fact that many of the loudest voices in the "war on fat" stand to gain money or authority from people's belief that they can and should try to lose weight. For example, Laura Fraser claims, "Diet and pharmaceutical companies influence every step along the way of the scientific process. They pay for the ads that keep obesity journals publishing. They underwrite medical conferences, flying physicians around the country expense-free and paying them large lecture fees to attend."[85] She also documents the financial ties between many of the lead researchers on studies that claimed that obesity was a dangerous disease and diet drugs or weight-loss clinics.

These conflicts of interest may even have influenced the current medical definitions of "overweight" and "obesity." The International Obesity Task Force within the World Health Organization, which recommended in 1987 that the BMI ranges for both categories be lowered, received funding from pharmaceutical companies Hoffmann-La Roche and Abbott Laboratories, which manufacture the weight-loss drugs Xenical and Meridia. The nutritionist who chaired the task force, Philip James, was paid to conduct clinical trials on both drugs.[86] Studies cited in the report that led to the change in the definitions actually showed no statistically significant correlation between premature mortality and a BMI below 40. Nevertheless, both the WHO and the National Institutes of Health adopted new BMI standards of 25 for "overweight" and 30 for "obese" in 1988, making 37 million Americans who had been a "healthy" weight the day before "overweight" overnight and subject to official medical recommendations that they lose weight to improve their health.[87] Of course, that also made many more people eligible for Xenical and Meridia prescriptions. According to Jeffrey Sobal's summary of how fatness became a disease, "The medicalization of obesity as a process did not occur by chance, but was a process which gained momentum as medical people and their allies made increasingly frequent, powerful and persuasive claims that they should exercise social control over fatness in contemporary society."[88]

The theory that the "obesity epidemic" and rise of weight-loss dieting were manufactured by the diet and pharmaceutical industries resembles the hypodermic needle model of communication, in which an all-powerful media injects messages directly into the audience's head. That model is generally considered obsolete, as media effects research has found that public opinion and behavior have a much more subtle and complicated relationship to mass media. Audiences do not simply passively accept whatever media messages tell them, even when the messages are clear and direct. Particularly in narrative genres such as film and television, the message is usually open to interpretation. Neither the WHO and the NIH nor *The Biggest Loser* invented

the meritocracy of thinness and injected it into Americans' heads, just as
Ratatouille did not invent the idea that high-class French restaurants serve
superior food that takes exceptional talent and skill to produce. These stories
would not be as popular as they are if they did not resonate with how many
Americans already see fatness and gourmet food. However, that doesn't mean
they merely reflect the status quo. Mass media may reaffirm or strengthen
viewers' faith in the dominant ideology of meritocracy. They do this not by
portraying the meritocracies of taste and thinness as perfectly reliable but
instead by dramatizing the contradictions to the dominant ideology and
providing audiences with the opportunity to embrace meritocracy in spite of
those contradictions.

People who think calorie restriction dieting can make anyone thin haven't
necessarily been duped by the diet industry or mass media weight-loss
narratives such as *The Biggest Loser*. Instead, audiences actively choose to
dismiss the contradictions *The Biggest Loser* portrays and, often, their own
experiences of diet failure. The meritocracy of thinness provides the narrative framework through which they interpret both the show and their own
experiences. Even people who personally struggle to achieve a healthy weight
often embrace the idea that they are responsible for their body size. Rather
than seeing weight regain as evidence that dieting fails, they tend to blame
themselves for their failure and assume that dieting works if done correctly.
The same desire to affirm the democratic, can-do spirit that audiences loved
in *Ratatouille* drives the popularity of *The Biggest Loser*. In their eagerness to
believe that they have access to status goods such as good taste and thinness,
Americans actively work to interpret mass media stories about food as being
even more inclusive and optimistic about the universal potential to achieve
the markers of class distinction than the stories themselves seem to be.

CHAPTER 5

Just Mustard

NEGOTIATING WITH FOOD SNOBBERY

*Language is terribly moralistic. . . . To call a man a snob, for instance, is a
very vague description but a very clear insult.*

—George Santayana

The previous chapters have focused primarily on the dominant cultural logic
of the food revolution; that is, the idea that aspirational foods such as expensive wine and organic vegetables are truly better than Budweiser and conventional produce and therefore the recent rise in the popularity of the former
is the result of a mass culinary enlightenment. Most mass media about food
produced since the late 1970s is devoted to helping people learn about these
supposedly superior foods—explaining why they're so much better and how
to find, prepare, and appreciate them. However, there is an also undercurrent
of doubt, and one of the symptoms of that doubt is a pervasive anxiety about
food snobbery. From pundits calling Obama's comments about arugula elitist
to the portrayal of a sneering restaurant critic as the villain in an animated
film, the popular discourse of the food revolution has been dogged by the
notion that liking or even just knowing something about superior food is a
form of class pretension. The idea of food snobbery implicitly challenges the
idea that eating better is a rational preference, threatening to unmask whole
food revolution project as merely a kind of class gatekeeping.

As George Santayana says, the words "snob" and "snobbery" can be
somewhat vague. Their multiple meanings reflect something of the history
and nature of social class divisions in the Anglophone world. "Snob," which
entered colloquial use in the late eighteenth century, was initially used to
refer to the ordinary classes or people without titles.[1] At least five possible
etymologies have been proposed: it may have been adopted from an abbrevia-
tion of the Latin *sine nobilitiate* (an official designation for aristocrats), the

Scottish word for cobblers and their apprentices (i.e., middling tradesmen), the French elision of *c'est noble* (it's noble), a Scandinavian word meaning charlatan, or a negation of the British schoolboy slang for nobles, or "nob."[2] Perhaps referring to ordinary people by a Latin or French designation for aristocrats was ironic, or maybe referring to all untitled people as cobblers was some kind of insult. What's clear is that from the beginning, the word "snob" had something to do with status, and it wasn't a compliment. Its emergence around the time of the Industrial Revolution seems likely to have been a response to the disruptions in the social order caused by the eclipse of Britain's landed nobility by non-aristocrats—factory owners, merchants, and the emerging professional-managerial class.

By the mid-nineteenth century, when William Thackeray published a collection of satirical profiles called *The Book of Snobs*, the term had narrowed to refer specifically to ordinary people who imitate nobles. The practice of emulating the rich was nothing new. Sumptuary laws designed to prevent commoners from dressing like aristocrats and make sure disfavored groups such as prostitutes were clearly identified date at least to the seventh century B.C. in Greece. By the thirteenth century, English law specified what colors and types of fabric and trim persons of various ranks or incomes were allowed to wear.[3] The colonial laws Massachusetts Bay passed in 1651 imposed a fine of ten shillings on anyone with personal fortune of less than £200 who dared to wear gold or silver lace, gold or silver buttons, or silk hoods or scarves.[4] However, these laws were often poorly enforced and openly flaunted, especially after the Industrial Revolution made it much easier for more people to acquire goods that had previously been out of reach for all but the aristocracy. As the dismantling of the old class system and the rise of the new bourgeoisie continued, snobbery began to acquire a second meaning: someone who looks down on people of a lower social status. The term also began to lose its initial association with commoners, or people without titles.

The result is that today the word "snob" refers both to ordinary people who commit the sin of "getting above their raising" and to anyone—rich or not— who sneers at people they deem inferior. Either of these meanings might be responsible for the word's pejorative nature and its association with a sense of hypocrisy. Aping the rich might be hypocritical on its own, because someone emulating a class cannot really belong to that class and the act of emulation implicitly slights the class they do belong to. But there's another level of hypocrisy, too. By acknowledging that class status is at least in part about performance rather than some sort of essence, snobbery undermines status hierarchies. The very idea of snobbery exposes the instability and socially constructed nature of class distinctions. It reminds people that, theoretically at least, anyone could pass as richer or more prestigious than they are, so the signs of class status might be misleading. At the same time, the act that "snobbery" refers to reinforces class hierarchies by investing in the social value of the signs that communicate class status. Emulating the rich suggests that it really is better to have and do the things the elite has and does and that class hierarchies are not just arbitrary social constructions. Furthermore, it attests to the risks of trying to pass and failing: the snob is a fraud who doesn't really belong. As an insult, the word conjures up the possibility of class transgression and punishes the failure to exploit that possibility well enough in one breath.

Even when referring to tastes that don't have a simple, one-to-one relationship to class status, snobbery typically carries negative connotations. A music snob might not listen to music associated with the rich, but their preferences must be odd or somehow antagonistic toward the popular, which runs afoul of the populist strain in American culture. On the other hand, ambition is often applauded, and many of the manners and tastes associated with higher class status are constructed as virtuous and meritorious. I suspect that the reason food snobbery poses such a potent threat to the food revolution isn't just because snobbery is inherently hypocritical or because it implies a disdain for the masses but also because it exposes widespread doubts about whether the foods that have been constructed as superior really are. If everyone were entirely, 100 percent convinced that their preferences for gourmet, diet, natural, and ethnic foods were justified by objective differences in taste, health, sustainability, or authenticity, there would be no fear of food snobbery. Anxieties about food snobbery are possible only because people realize that participation in the food revolution could be motivated by concerns about status. To prove that status concerns aren't the real motive for participating in the food revolution, people have developed a diverse set of strategies for negotiating with the threat of being outed as class-climbers.

This chapter examines three different strategies for combating that threat. The Rolls-Royce commercials for Grey Poupon mustard and their many parodies seem to associate the condiment with the elite, but they actually undermine the idea that enjoying fancy mustard is elitist. As a parody of the rich, the Rolls-Royce commercials suggest that food is too mundane to represent real snobbery. The critically acclaimed film *Sideways*, which also became a surprise box-office hit, creates a distinction between snobbery and what I call connoisseurship. It portrays the former, performing tastes to impress other people, as ignoble but the latter as admirable and suggests that it can be demonstrated through the selective appreciation of low-status foods. Finally, the term "foodie" and the backlash against it exemplifies the strategy of discarding trends once they become tainted with elitism, similar to the process by which trends lose their cachet once they become too common.

Denial is not an admission of guilt. The strategies people use to deflect the suspicion that their tastes amount to snobbery don't prove that people really are eating better primarily because of concerns about status. However, at the very least they expose the threat of class anxiety that runs through the popular discourse about food and taste. They also demonstrate that contrary to what some people claim, Americans are aware of social class differences and they do talk about them. Snobbery has always turned on questions of what class someone belongs to and how they regard people in other classes. Concerns about food snobbery amount to a widespread, tacit acknowledgment of how classes in the United States today are demarcated by taste. Furthermore, the pervasiveness of the anxiety about snobbery suggests that there's more going on here than simply a conflict between populism and class aspiration. I suspect that the tenacity of the threat of food snobbery reflects the guilty conscience of people who know, on some level, that their attempts to eat better might have more to do with status than substance.

JUST MUSTARD

The phrase, "Pardon me, but would you have any Grey Poupon?" was introduced into the American lexicon by a Madison Avenue ad man named, improbably, Larry Elegant. Elegant was a copywriter for Lowe Marschalk, the agency the Heublein company hired in 1980 to create a television commercial to supplement a new print and product-placement campaign for their then-unknown brand of Dijon-style mustard. The 30-second spot that Elegant came up with opens with the backseat passenger's view of a car being driven by a man in a classic chauffeur's uniform while baroque orchestral music plays. A voice-over with a British accent says "The finer things in life. Happily, some of them are affordable" as the chauffeur opens the glove-compartment box to reveal a jar of Grey Poupon.[5]

The camera swivels to follow the chauffeur's gloved hand as he gives the jar to the backseat passenger, a man wearing a suit and a tie. A small table in front of him is set with cloth linens, china plates with sliced meat and a green salad, silver utensils, and glass stemware. Images of the passenger eating are interspersed with glamour shots of mustard being spooned onto other elegant plates and a golden-hued dressing being poured over salad. The voice-over continues, saying that Dijon mustard is "so fine it's even made with white wine," and then listing a variety of applications, ending with "and of course, sandwiches." The video cuts to a shot of the car from the outside as it comes to a stop. A second car with the same distinctive Rolls-Royce grille pulls up alongside it, and the second car's passenger, another man in suit and tie, leans toward the open car window and delivers the famous query. "But of course," the first passenger replies, extending the Grey Poupon out the window. The camera zooms in on the jar and the shot freezes as it passes between their hands. Beneath the jar, the tagline "One of life's finer pleasures" appears on the screen.[6]

In the handful of East Coast cities where the commercial first aired, sales jumped 40 to 50 percent, prompting Heublein to go national with the campaign.[7] The "Pardon me" ads have since been widely credited with catalyzing America's shift from a yellow mustard monoculture to a country where even gas-station Subways have giant plastic squeeze bottles of Dijon mustard along with all their other condiment choices. The campaign's success has made it something of a legend in modern marketing, even though its overt strategy is hardly remarkable. The advertising historian Roland Marchand notes that even in the earliest forms of national advertising, "the most obvious source of distortion in advertising's mirror was the presumption by advertisers that the public preferred an image of 'life as it ought to be.'" According to Marchand, even during the Great Depression, "ad creators tried to reflect public aspirations rather than contemporary circumstances" and "often sought to give products a 'class image' by placing them in what advertising jargon would call 'upscale' settings."[8]

Elegant himself told Malcolm Gladwell, who wrote about the campaign in a 2004 New Yorker article, that creating a "class image" for Grey Poupon was essentially how the "Pardon me" ad worked: "The tagline in the commercial was that this was one of life's finer pleasures, and that, along with the Rolls-Royce, seemed to impart to people's minds that this was something truly different and superior."[9] Advertising-industry trade publications seem to agree. In 1984, Madison Avenue reported that Grey Poupon had captured 80 percent of the U.S. market for Dijon mustard and attributed its success to three factors: "1. the popularity of nouvelle cuisine, 2. impressive growth of the whole Dijon category, and 3. an upscale, national television advertising campaign that successfully linked mustard with opulence."[10]

However, product development and marketing history is full of examples where the same strategy failed and few where it succeeded so spectacularly. In the same New Yorker article, Gladwell recounts ketchup entrepreneur Jim Wigon's struggle to get people to buy his upscale version of mustard's most common counterpart. According to Gladwell, Wigon's story challenges the notion that the "Pardon me" ads deserve primary credit for Grey Poupon's remarkable success. While Gladwell admits that the sophistication imparted

by the ads may have gotten some people to try the mustard, he argues that
Grey Poupon ultimately succeeded because it simply tastes better than yellow mustard: "One day the Heublein Company, which owned Grey Poupon,
discovered something remarkable: if you gave people a mustard taste test, a
significant number had only to try Grey Poupon once to switch from yellow
mustard. In the food world that almost never happens; even among the most
successful food brands, only about one in a hundred have that kind of con-
version rate."[11] He argues that this explains why concerted efforts on the part
of people such as Wigon to fill what seems like a gaping hole in the condiment market have failed. Heinz is already as palate pleasing as it is possible
for ketchup to be. The implication is that people won't buy whatever you're
selling just because you put it in a Rolls-Royce and tell them it's made with
wine—it must actually taste better.

Gladwell's explanation relies on the logic of culinary enlightenment; it
presumes that if certain foods succeed and others fail even when they're
marketed in the same way, then the successful ones must be objectively
superior. However, Gladwell's initial description of Jim Wigon's "World's Best
Ketchup" suggests that it, too, might have a gustatory advantage over Heinz.
After spending a day watching people taste free samples of Wigon's Ketchup
at Zabar's specialty food store in New York, he reported, "The ratio of tomato
solids to liquid in World's Best is much higher than in Heinz, and the maple
syrup gives it an unmistakable sweet kick. Invariably, people would close their
eyes, just for a moment, and do a subtle double take."[12] Ninety people who
tasted it that day liked it well enough to purchase a jar. Still, Gladwell suggested that Wigon's failure to capture more of the condiment market ultimately comes down to the superiority of Heinz. He spoke to several flavor
experts who reported that Heinz has a high "amplitude," meaning its different
components are well balanced, a characteristic many other iconic brands,
such as Hellman's mayonnaise, Sara Lee pound cake, and Coca-Cola, share.
In comparison, a panel of trained tasters at Kansas State University said
Wigon's ketchup tasted, inexplicably, "more like a sauce."

But condiment purchases are not a zero-sum game. French's classic yellow is still America's best-selling mustard, and many people now keep both
yellow mustard and Dijon in their refrigerators. It remains unclear why the
"complex aromatics" of Grey Poupon would catapult it from a niche brand
that grossed only $100,000 annually to one of the most recognized names
in condiments, but the double-take-inspiring flavor of Wigon's ketchup (or
some other, less saucy gourmet version) has yet to carve out a similar space
in the supermarket condiment aisle. Gladwell acknowledges that for most
food products, diversification is the name of the game. In the 1990s, Ragu

dramatically increased its spaghetti sauce sales by diversifying beyond its classic formula. Apparently for some foods, what people want is not necessarily the single highest-amplitude version, but a choice between smooth and chunky, sweet and zesty.

Maybe ketchup is just different, but the theory that "high amplitude" or objectively superior flavor is what makes brands such as Grey Poupon and Hellman's successful is undermined somewhat by devotees of even more expensive and exclusive brands of mustard, such as Maille,[13] or who insist on the superiority of homemade mayonnaise. Furthermore, even if superior flavor were the key to the rise of Dijon, that doesn't explain why the Rolls-Royce commercials themselves became a cultural touchstone.

Even before trade publications such as *Madison Avenue* were heralding Grey Poupon's sales success, the mustard had become comedic fodder. In the 1980s, allusions to the mustard appeared on *The Tonight Show*, *The Fresh Prince of Bel-Air*, *Married with Children*, *The Simpsons*, and *Mad TV*, always in the context of caricatures of wealth and luxury. In the hit 1992 film *Wayne's World*, the Gen X metalhead protagonists reenact the commercial when they pull up next to a Rolls-Royce stopped at a traffic light.[14] In 2007, almost three decades after the original commercial aired, it was referenced again in the *Star Wars*–themed season premiere of the animated series *Family Guy*. In the midst of an extended car chase, two sandcrawlers pause at a stoplight, roll down their windows, and hand off a jar of Grey Poupon.[15] According to a list of the "100 greatest television catchphrases" published by the cable network TVLand in 2006, "Pardon me sir, would you have any Grey Poupon?" ranked #69, just ahead of "Marcia, Marcia, Marcia!"[16]

Some allusions simply invoke the brand name, although even then its cultural resonance is likely to be due to the Rolls-Royce campaign. However, many specifically reference elements from the commercials. The 2006 single "We in Here" by DMX includes the lyrics, "if I pull up on, it won't be for Grey Poupon."[17] In the song "Show It to Me" on T.I.' s 2007 *T.I. vs. T.I.P*, featured guest Nelly says, "Yea you rollin' wit the King and the one / Country black folk in the Chevy passin' Grey Poupon."[18] Dozens of user-produced images on the website *I Can Has Cheezburger?* are emblazoned with either the original query or an obvious play on it, and many also visually reference the Rolls-Royce scene.[19] Additionally, if the comments on the YouTube video of the original commercial are to be believed, dozens of people have re-created the "Pardon me" scene on their own.[20] The popular remixing of the ad's tag line calls into question the assumption that the campaign's success owes entirely, or perhaps even primarily, to a successful association between condiment and sophistication.

Elegant and *Madison Avenue* consider only Grey Poupon's sales success when they assume that the ad worked as intended, by cementing a relationship between the brand and the idea of culinary superiority in the minds of consumers. However, the proliferation of parodies suggests instead that the commercial's appeal was at least partially in its critique of class pretensions. "Pardon me" became a catchphrase because it was an effective caricature. The ad resonated widely as a mockery of ostentatious wealth. Later versions of the ad also testify that something other than Marchand's idea of the "class image" is at work. In 2007, a spokesperson for Kraft Foods, which acquired the brand in 1999, told *Advertising Age* that they had decided to revive the Roll-Royce because consumers "know, love, and associate the 'Pardon me' campaign so strongly with the Grey Poupon brand." The new 30-second spot was almost exactly the same as the original, with similar music and a similarly attired passenger in the back seat of a chauffeur-driven Rolls-Royce enjoying a virtually identical meal. Shots of vinaigrette being poured over salad and a dollop of mustard being spooned onto a plate of sliced meat also clearly reference the original ad. However, instead of handing over the jar when the second car's passenger asks, "Would you have any Grey Poupon?" the passenger with the mustard simply says, "But of course," and then his driver pulls away, leaving the second passenger flustered and mustardless.[21] By mocking the heightened formality of the "would you" syntax, the 2007 ad lampooned both the 1980 commercial and the upper class supposedly associated with its product.

This was not the first time the campaign had poked fun at itself. In 1995, when then-owner Nabisco introduced a new squeeze bottle, they ran a commercial showing a man in the back seat of a Rolls-Royce assembling a sandwich. As he dispenses the mustard, an air bubble bursts in the bottle's opening, making a farting noise. The chauffeur raises his eyebrows, and the passenger exclaims, "Pardon me!"[22] Although the 1997 and 2005 commercials were explicitly arch, even the original was predominantly viewed ironically, whether or not that was Elegant's intention. Instead of reinforcing the idea that anything other than yellow mustard was a snooty luxury, the commercial highlighted how ludicrous it would be to believe that something as mundane as a condiment could be a marker of privilege. The idea of keeping a glass jar of mustard in a Rolls-Royce glove-compartment box, having a multicourse dinner with wine in the back seat of a car, or sharing a condiment jar between chauffeured vehicles shot directly past the "image of 'life as it ought to be'" and landed directly in *Saturday Night Live*'s territory. Notably, many of the parodies didn't significantly alter the content of the commercial—they simply reenacted the scene because it was already taken as a joke. It's possible that by portraying the brand as being in on the joke, the ads actually brought

Grey Poupon over to the side of the average consumer, aligning it against the lampooned limousine set.

The original commercial also made claims about Grey Poupon's ordinariness that undermined the signs of extraordinariness such as the limousine and fancy meal. The list of possible applications and the reassurance that Grey Poupon could "of course" be used on sandwiches implied that it could substitute for good old yellow mustard. Despite everything about Grey Poupon that might have seemed pretentious, such as its French appellation, the glass jar, and the wine in the ingredient list, the commercial worked to convey the idea that it was still just mustard. That isn't to say that people didn't perceive Grey Poupon as a classier product than French's. Instead, the ad succeeded because it negotiated the contradiction between the desire to enjoy "life's finer pleasures" and the stigma about class-climbing the word snobbery polices. Widely seen as parody of the rich, the "Pardon me" commercial offered consumers plausible deniability that buying a gourmet mustard was a sign of elitism. For nearly thirty years, the campaign effectively reinforced the idea that Grey Poupon is an accessible, unpretentious luxury by ridiculing the idea that a condiment could be truly snobbish.

SIDEWAYS AND THE VALUE OF CULINARY CAPITAL

Like the Grey Poupon commercials, the 2005 film *Sideways* works to reassure audiences that appreciating luxury foods—or in this case, the quintessential luxury beverage, wine—is not inherently snobbish. However, *Sideways* performs a more complex negotiation with snobbery that preserves the value of high-status foods. Instead of portraying wine as too mundane to be snobbish, *Sideways* suggests that both ignorance about wine and wine snobbery are bad. It elevates a third alternative: an appreciation for wine based on its inherent characteristics rather than its status connotations, which I'll call connoisseurship. It also suggests that one way the connoisseur can distinguish herself from the snob is by demonstrating an appreciation for low-status foods. A selective appreciation of low-status foods affirms that someone's upscale tastes are genuine preferences, not just a way of aping the rich or performing culinary capital.

In the essay where Kathleen LeBesco and Peter Naccarato introduce the term "culinary capital," they depart from Pierre Bourdieu's concept of cultural capital when they argue that the symbolic capital Julia Child and Martha Stewart offer their fans is "illusory." According to Bourdieu, cultural capital is a real form of capital that is "convertible, on certain conditions, into economic capital."[23] LeBesco and Naccarato, in contrast, suggest that the performance of a desired class identity through high-status foods merely creates the illusion of mobility. The consumer who buys products in Martha Stewart's exclusive Kmart line might think she's making her home look more like the upper-middle-class homes disproportionately represented in prime-time television programming, but she's wrong. The fact that she's shopping at Kmart reveals the truth of her class status.

It's true that watching lifestyle programming is unlikely to boost most people into a higher income bracket, even if the knowledge they acquire enables them to learn more about the tastes of the rich and sometimes emulate them. Nonetheless, given that culinary capital, like all forms of cultural capital, is socially constructed, it's unclear what would distinguish food and eating practices with real value from the "illusions" offered by the likes of Child and Stewart.[24] I think Bourdieu had it right: culinary capital is a source of real value even if it usually isn't enough by itself to change someone's class status. What LeBesco and Naccarato seem to be identifying when they claim that foodways are "vehicles for performing an illusory identity" is not actually that the desired identity is illusory but instead that the idea that everyone has access to that identity is a myth. In other words, the problem is not that culinary capital is fake capital; the problem is that the meritocratic ideology that drives the popular appeal of lifestyle programming and a lot of foodcentric mass media is unreliable.[25] Talent and hard work aren't always rewarded. Being able to perform the gestures of wealth won't necessarily make you rich. However, the fact that the meritocracy of taste is mostly mythical doesn't make culinary capital any more illusory than the fact that meritocracies of occupational success and wealth make material capital illusory. There are real rewards for demonstrating high-status culinary capital in the right contexts, and there are real costs for failing to do so.

One of the opening scenes of *Sideways* gestures toward the costs of failing to demonstrate the right culinary capital. The film was adapted from a novel by Rex Pickett about a middle-aged wine aficionado named Miles who takes an old college buddy named Jack on a road trip to the Santa Ynez Valley, where some of the most expensive and sought-after wines in California are produced. The trip is a kind of bachelor's party for Jack, and as the men head away from Los Angeles in Miles's Saab convertible, Jack accuses Miles of being late to pick him up from his future in-laws' house because he was hungover. Miles concedes, "Okay, there was a tasting last night, yes, but I wanted to get us something nice for the ride up." "Check out the box," he says, gesturing toward the back seat. Jack selects a bottle with the caged cork indicative of a sparkling wine and begins to open it despite Miles's protests that he shouldn't open that one because it's not chilled. As the wine erupts from the bottle, only mostly into the stemware that Jack seems to produce out of nowhere, Miles laments, "Half of it . . . gone!" Jack hands him the first glass—the more appropriate champagne flute—and then pours one of his own—a regular wine glass. "Hey, shut up, okay?" Jack says, "Here's to a great week. Come on." Miles sighs and then nods, clinking Jack's raised glass and saying, "Yes, absolutely. Despite your crass behavior, I'm actually glad we're getting this time together." Jack takes a sip and then does a double take: "Man, that's tasty." Miles nods and says, "That's 100 percent Pinot Noir, single vineyard. They don't even make it anymore." Jack looks at his glass again quizzically, and says, "Pinot Noir . . . then how come it's white?" "Oh, Jesus," Miles scoffs. "Don't ask questions like that up in wine country. They'll think you're some kind of dumb shit, okay?"[26]

The majority of the audience, presumably not wine experts, might identify with Jack's confusion about the fact that Pinot Noir grapes can produce a golden, effervescent wine. Miles's derision is likely to seem pompous, but several elements in the scene suggest that he really does have superior knowledge and taste. The fact that the wine spills all over the car proves that his advice about not opening the unchilled bottle wasn't just a fussy preference. Despite the suboptimal temperature, even Jack recognizes that the wine is especially tasty. Additionally, Miles provides a clear (and accurate) explanation of the difference between red and white wines, suggesting that there is an objective basis for differences between wines that can be learned and that Miles has cultivated that knowledge. Miles may be pretentious, but he's not just pretending when it comes to wine.[27] The scene also demonstrates how even relatively private acts of eating and drinking, as Miles and Jack undoubtedly assume that no one else can see them drinking alcohol from open containers in a moving vehicle, are nonetheless opportunities to acquire and rehearse culinary capital.

There's more at stake in the acquisition of culinary capital than just enhancing your personal drinking experiences. Miles's concern about what Jack's ignorance might make people in wine country think about them reflects an awareness of how taste influences social judgments. However, that awareness turns out to be Miles's primary weakness. His knowledge of wine may be real and valuable in some social contexts and his enjoyment of it certainly seems genuine enough, but his concern about what other people think about his taste is at odds with the ideal of disinterested connoisseurship. *Sideways* suggests that in order to avoid being a snob, you must not only possess the right culinary capital, you must also appreciate food regardless of its relationship to status hierarchies. The only acceptable deployment of culinary capital depends on the renouncement of its value.

The Rube, the Snob, and the Connoisseur
The Rube

Despite Miles's best efforts to educate Jack, once they get to wine country, he does turn out to be a dumb shit. At first, Jack seems to be a proxy for the audience, modeling both their likely unfamiliarity with wine-tasting culture and their bemusement at Miles's fussiness. In their first visit to a tasting room, Miles explains the full ritual: holding the glass up to the light, tipping the glass to evaluate the color and opacity, sniffing, swirling, and sniffing again. Jack follows along dutifully, then asks the question the audience may also be wondering at that point, "When do we drink?"[28] However, the film continues to discourage the audience from identifying too much with Jack. His ignorance is the butt of many of jokes, and in most cases the viewer gets to be in on the joke by virtue of knowing more than him. Even viewers who don't know why people swirl the wine before drinking it will probably know that it's unseemly to down a tasting pour like a shot of cheap tequila (see Fig. 14). Worse, as they're leaving the first tasting room, Miles notices that Jack was chewing gum the entire time they were there.

The audience is further discouraged from sympathizing with Jack by his general boorishness. Although Jack confidently introduces himself as an actor, it turns out that aside from a bit part on a daytime soap opera a long time ago, the only work he can get is reading the disclaimers at the end of commercials. At breakfast the first morning after their arrival in wine country, Jack leers at their young waitress and declares that his best man's gift to Miles

Figure 14. Jack gulping the tasting pour while Miles sips behind him. Source: *Sideways*, dir. Alexander Payne, DVD, Fox Searchlight Pictures, 2004.

is going to be to get him laid, despite Miles's protests that he'd rather have a knife.[29] The next morning, when Miles lays out a carefully planned itinerary featuring another series of visits to vineyard tasting rooms, Jack explodes: "I am going to get my nut on this trip, Miles. And you are not going to fuck it up for me with all your depression and anxiety and neg-head downer shit. . . . I am going to get laid before I settle down on Saturday. Do you read me?"[30] Miles's plan, and indeed the whole trip, may be selfishly oriented toward his own interests, but Jack's insistence on spending his bachelor party cheating on his fiancée is likely to strike many viewers as even less admirable.

Jack quickly achieves his goal, picking up a sultry wine pourer named Stephanie at one of the vineyards they visit and neglecting to tell her about his upcoming marriage. Miles is shown golfing alone and being shooed out of the hotel room while Jack and Stephanie have raucous sex and then pal around with her kid. After Stephanie finds out about Jack's engagement, she breaks his nose with her motorcycle helmet in a fit of rage. Even after that, Jack remains undeterred. He starts flirting with the first waitress they encounter after leaving the emergency room and clearly thinks it's his charm that convinces her to take him home. However, it turns out that he's been suckered into some kind of cuckold fantasy scam. The waitress's husband comes home to find them mid-coitus, as the couple has clearly prearranged, and he chases Jack out of the house without his clothes or wallet. Jack cajoles Miles into sneaking back into the house to retrieve his wallet, which contains his wedding rings. Then, without warning, Jack drives Miles's car into a tree to buttress the lie he plans to tell his fiancée about the origins of his bandaged nose. Jack remains a blundering jerk who thinks only about himself to the end of the film, and his ignorance about wine and indifference to the rituals of fine dining are aligned with his buffoonery.

The Snob

At least as far as *Sideways* is concerned, Miles with his superior taste isn't much better than Jack and his culinary ignorance. During the opening credits, Miles goes into a coffee shop to order a triple espresso, the *New York Times*, and spinach croissant, which he pronounces with a French accent: "*kwa*-san" rather than the Americanized "kreh-*sont*." The affected pronunciation and the fact that he specifically asks for the *New York Times* rather than just grabbing one off the rack (and in Los Angeles, no less) are not only stereotypically elitist but also highly performative; he's not just seeking out the refined pleasures he happens to enjoy, he's obviously trying to impress the people around him with his superior taste, even if the behaviors are so

Figure 15. Miles detecting "just a flutter of uh, like a, nutty Edam cheese" while Jack watches incredulously. Source: *Sideways*, dir. Alexander Payne, DVD, Fox Searchlight Pictures, 2004.

habitual he doesn't even realize he's doing it. Later, his histrionic insistence that he won't drink "any fucking Merlot," seemingly because it's too popular, is clearly exaggerated for the sake of humor. His insistence that he smells "the faintest soupçon of, like, asparagus and just a flutter of a, like a, nutty Edam cheese" in the wine he uses to model the rituals of tasting for Jack is also played for laughs, just like Jack's gum chewing.[31] In contrast with the shot of Jack downing the tasting pour, as Miles squints with concentration to discern those subtle (or imagined) flavors, the camera focuses on Jack, who models the audience's probable incredulity (see Fig. 15).

Further undermining Miles's credibility, *Sideways* also suggests that his taste for the finer things isn't a reflection of his income or upbringing. He is a divorced middle-school English teacher with a modest apartment, and when they stop at his mother's house in the suburbs of Los Angeles on their way out of town, ostensibly to wish her a happy birthday, it's apparent that he didn't grow up rich. She lives in a small condo with mismatched furnishings and tacky, dated bric-a-brac on the walls. He promises Jack they won't stay long, but when she offers them food he immediately says, "Yeah, I'm hungry," and then feigns helplessness as she insists they stay the night. Jack offers a back-handed compliment about the dinner she serves them on mismatched plates: "This is delicious Mrs. Raymond, absolutely delicious. . . . Is this chicken?" which further establishes that Miles didn't get his epicurean tendencies from her. Miles's real purpose for the detour is revealed when he sneaks upstairs to her bedroom and roots around in her dresser until he finds a can of Ajax. He expertly twists off the bottom and a roll of $100 bills wrapped in rubber bands slides out. He peels off at least $1,000 before rewrapping the bills and returning them to the can. His shame and self-loathing is palpable, especially after he returns to the dinner table and his mother asks in an indiscreet whisper, "Do you need some money?"[32] The fact that Miles has to steal from his mother to afford the vacation not only makes him an archetypal loser, it also establishes that his sophisticated tastes are a form of pompous posturing that he can't really afford.

Like Jack's ignorance about food, Miles's snobbery corresponds with his other character flaws. The lies he tells to Jack and his mother about the reason for their detour to her house are of a piece with his constant attempts to use his wine smarts to delude everyone, including himself, about how pathetic his life is. Instead of a refined hobby, his love of wine is largely a flimsy disguise for his use of alcohol to escape from his failures, a dependence that might border on alcoholism. He insists that they walk to distant vineyards so as not to have to "hold back" and plans their tasting room visits so "the more [they] drink the closer [they] get to the motel."[33] After Jack accidentally lets it slip that Miles's ex-wife has remarried, Miles guzzles an entire bottle as Jack chases him through a vineyard. At dinner with Stephanie and her friend Maya, Miles drinks so much that he slips into a dark mood and excuses himself to drunk dial his ex-wife from the restaurant pay phone and passive-aggressively slur at her that she doesn't have to worry about running into him at Jack's wedding because he's decided not to go. Her initial concern about why he's calling her so late fades quickly into a weary, "Oh Miles. You're drunk," suggesting that this is a habitual behavior and likely played a role in their divorce.[34]

In one telling scene, Miles gets so desperate for a drink that he momentarily drops the pretense of pickiness. Midway through a visit to big commercial vineyard that Jack insisted they stop at, after already declaring that their wine tastes like "rancid tar and turpentine bullshit," Miles receives a voicemail alert. He walks into the parking lot to listen to the message, which turns out to be from his literary agent. She says that the publisher who had been interested in his novel has decided to pass and that she doesn't think she'll be able to find another one. Devastated, Miles marches back into the tasting room, steps up to the bar, and demands a pour. He downs it in one gulp, just like Jack did at the first tasting room, replaces his glass on the bar and says, "Hit me again." He downs the second pour just as quickly and impatiently asks the pourer for a full glass, offering to pay. When the man refuses, suggesting that he buy a bottle and go drink it in the parking lot, Miles grabs the bottle out of the man's hand and fills his glass nearly to the brim. The pourer grabs the glass and they struggle over it, spilling the wine in the process. Miles steps back, momentarily defeated, but then glances at the bucket sitting on the bar, full of the expectorated tastings of dozens of strangers. He grabs it, and pours it into his mouth and all over his face and shirt, making the other people in the tasting room groan in disgust.[35]

Instead of being portrayed as a form of enviable sophistication, Miles's taste is portrayed as ridiculous pedantry at best and an ugly addiction at worst. His drunken escapades drive as many of the film's moments of outlandish farce as Jack's libido. The apparent contrast between the two characters that was established in the opening scenes in the car turns out to be a red herring. Jack turns out to be less a foil than a double for Miles. They are both selfish losers who lie to and hurt the women in their lives and delude themselves about who they really are. In the tasting room scene, Jack and Miles are framed the same way. When Jack is downing the pour, he is in the foreground but the focus is on Miles behind him; when Miles is concentrating on the subtle aromas in the glass, he is in the foreground with the focus on Jack behind him. This visual symmetry reflects the similarity between the characters. Miles's posturing is portrayed as just as ludicrous and pathetic as Jack's boorishness. The difference is that whereas Jack remains the same to the end, Miles is allowed to evolve. In fact, the resolution of the film depends on his redemption, which, like so many things in the film, is symbolized by wine.

The Connoisseur

The primary example in *Sideways* of a wine drinker who's neither ignorant nor pretentious is Maya, a waitress Miles has long had a crush on who turns out to be friends with Stephanie, Jack's wine-pourer paramour. Because of Jack and Stephanie's affair, Miles suddenly has the opportunity to spend time with Maya outside the restaurant where she works, and they exchange stories about their relationship to wine in a pair of monologues at the heart of the film. Although the scene appears to be a moment of touching connection between the characters—they're alone on a porch at night, which is warmly lit from the glow of the house, and they gaze into each other's eyes as they speak—their words actually offer a neat juxtaposition of their characters that is represented by their motivations for drinking wine.

First, Maya asks Miles to explain his fondness for Pinot Noir, which, as she notes, is "like a thing with [him]." He says he likes it because it's "thin-skinned, temperamental . . . not a survivor like Cabernet," but if carefully nurtured by someone who "really takes the time to understand its potential . . . its flavors, they're just the most haunting and brilliant and thrilling and subtle and . . . ancient on the planet."[36] The description is a thinly veiled portrait of Miles as he sees himself: unique, fragile, and often misunderstood but full of unrealized potential. After he trails off, he says, "What about you?" Whereas Maya's question was specific and reflected that she had paid attention to him, his is an afterthought. "What about me?" she asks to clarify, looking a little surprised that he even asked. "I don't know, why are you into wine?" he offers, shrugging. First, she credits her ex-husband, whom she says had a "big sort of show-off cellar." She continues:

> MAYA: Then, I discovered that I had a really sharp palate. And the more I drank, the more I liked what it made me think about.
> MILES: Like what?
> MAYA: Like what a fraud he was. [*Miles laughs and says "Wow" or "ow" uncomfortably.*] No, [*she laughs too*] I like to think about the life of wine, how it's a living thing. I like to think about what was going on the year the grapes were growing, how the sun was shining that summer or if it rained [. . .] all the people who tended and picked the grapes, and if it's an old wine how many of them must be dead by now [. . .] how every time I open a bottle it's going to taste different than if I had opened it on any other day because a bottle of wine is constantly evolving and gaining complexity, that is, 'til it peaks. . . . And it tastes so fucking good.[37]

Maya's love of wine is rooted in the wine itself—the grapes and the effort that went into growing them, the time represented by the aging process, and the taste. Unlike Miles, she doesn't single out or dismiss any particular varietal or seem remotely concerned about what anyone else thinks about the wine she likes. Although she acknowledges that every bottle has a peak, she talks about appreciating the particular taste of each bottle on the day you open it. Unlike both Miles and her ex-husband, Maya has no apparent interest in showing off. Her love of wine is based on her experience of wine, not what she hopes it will make other people think about her. Miles's laughter at her barbed comment about her ex is tinged with guilt because Maya's love of wine for wine's sake exposes him for the status-obsessed snob he is.

The film also implies that Maya has a better palate than Miles, or at least that her perceptions are less muddled by venal desires, like wanting to impress people or a need to just get drunk. Immediately before their monologues, she and Miles are talking in Stephanie's kitchen after having opened a bottle of wine. Miles takes a sip and immediately intones, "Wow, that's nice, that's really good." He swirls the glass and continues, "Need to give it a minute, but that's really tasty. How 'bout you?" Maya looks thoughtful and shakes her head: "I think they overdid it a little. Too much alcohol, it overwhelms the fruit." "Huh," Miles says and takes another sip. Then he praises her assessment: "Yeah, yeah, I'd say you were right on the money. Very good."[38] Her assessment refers to objective qualities such as the alcohol percentage rather than a "faint soupçon of asparagus."

Despite waxing poetic about why she loves wine, Maya turns out to be less sentimental about it than Miles. She immediately recognizes that the bottle he names as the prize in his collection, a 1961 Château Cheval Blanc, is peaking and urges him to drink it before it begins its decline. Instead of taking credit for this bit of expert knowledge, she immediately exhibits an appealing humility by noting that she "read that somewhere." When Miles says he's been waiting for a special occasion and that it was originally intended for his ten-year wedding anniversary, Maya says that the day you open a bottle like that, the wine itself is the occasion. Once again, Miles is hung up on what wine symbolizes while Maya advocates just enjoying it for its own sake.

In spite of Miles's many flaws, Maya seems to genuinely like him until he accidentally lets it slip that Jack is engaged. Complicit in his friend's lies to Stephanie, Miles seems to have blown his chance. However, unlike Jack, Miles seems chastised by the experience. Instead of making good on his drunken threat to avoid the wedding so as not to see his ex-wife, he dutifully fulfills his role as best man and is even cordial when she introduces him to her new husband in the parking lot after the ceremony. She tells him she's pregnant and he is clearly stung by the news but manages to congratulate her. Then, instead of following the other cars leaving the church parking lot, presumably in the direction of the reception, Miles drives in the other direction. Moments later, he parks hastily outside his apartment, runs up the stairs, and roots around at the bottom of a closet. The film cuts to a shot of a register at a fast-food restaurant. The camera slowly pans around the florescent-lit dining area where fat people shuffle around in sweat suits and zooms in on Miles, sitting alone in one of the vinyl booths still wearing his tuxedo shirt. He drinks from a large, lidless Styrofoam cup and there's a half-eaten burger and a pile of onion rings in front of him. He looks around surreptitiously to ensure that no one is watching and refills the cup from a bottle hidden in the corner of the booth.

The proper rituals he modeled for Jack, such as swirling the wine carefully in a stemmed glass are utterly abandoned. He actually has to hide the bottle from view and fill his cup surreptitiously instead of showing off what he's drinking. The slug lines in the script seem to recall Maya's monologue: "As the camera MOVES CLOSER, all the complex emotions inspired by the wine ripple across Miles's face."[39] The label is turned away from the camera as he pours, but a subsequent shot of it leaning against the corner of the booth

Figure 16. Surreptitiously drinking Château Cheval Blanc with a fast-food burger.
Source: Alexander Payne, dir., *Sideways*, Fox Searchlight Pictures, 2004.

confirms that it is the prize bottle he told Maya about, his '61 Château Cheval
Blanc (see Fig. 16).

The ultimate irony is that this particular wine is composed substantially
of the Merlot grapes he earlier claimed to despise.[40] While only oenophiles
are likely to pick up on that detail, it offers further evidence that the scene
represents a real departure for Miles, a break with his old, bad, snobbish self.
The Miles in the fast-food restaurant doesn't care if a wine happens to be
made of Merlot or about drinking from the correct glass or about impressing
anyone. He just enjoys how good the wine tastes. Even for viewers who don't
know anything about the particular blend of grapes in Château Cheval Blanc,
the setting and the meal are sufficient to communicate his transformation.

The fast-food burger, perhaps America's most prominent icon of populist, unpretentious food, cleanses Miles's love of wine of its unappealing elitism.

The scene in the burger joint is a turning point for Miles in other ways. After it ends, the words "five weeks later" appear and Miles is shown at the front of his middle-school English classroom. One of his students finishes reading a passage from *A Separate Peace* and Miles dismisses them for the weekend. He returns to his apartment to find a message on his answering machine from Maya. Her words, tentative but warm, play as a voice-over while Miles's convertible is shown driving onto the same highway ramp where he and Jack set out on in the beginning of the film and then climbing the steps to Maya's apartment. She says she received his letter and needed some time to think about it and that she also wanted to finish his novel, which he'd given her a copy of after their night on the porch. She calls it lovely, asks if he really went through all the "beautiful and painful" things in it, and says he should let her know if he'll be back in Santa Barbara anytime. As her message ends, he takes a deep breath and knocks. The scene in the fast-food restaurant paves the way for the possibility that buoys the ending: perhaps this new, redeemed Miles will finally succeed where the old, pretentious Miles was doomed to fail.

THE *SIDEWAYS* EFFECT

Sideways was widely hailed as a "surprise hit."[41] Its reception offers additional insight into how people come to terms with their acquisition of elite culinary capital despite the threat of snobbery. Fox Searchlight, the film's distributor, initially had difficulty getting theaters interested.[42] In its opening weekend, the film was shown in only four theaters and grossed a paltry $207,042. After it began attracting critical acclaim, including five Academy Award nominations and the Oscar for Best Adapted Screenplay, it was rereleased at 699 theaters nationwide. It ended its box-office run in May 2005 after grossing $71 million, which made it the 40th highest-grossing film of 2004 (out of 551 ranked by IMDb affiliate Box Office Mojo), ahead of many other films released that year with bigger budgets and more famous stars, such as Quentin Tarantino's *Kill Bill Vol. 2* and *The Stepford Wives* remake. Commercially, it also beat out several other successful independent films released that year, including *Eternal Sunshine of the Spotless Mind, Garden State,* and *Napoleon Dynamite.*[43] Its success is at least partially due to the writing, direction, and performances of the four lead actors, all of which received accolades and award nominations. However, the comments on Metacritic and the Internet Movie Database suggest that the focus on wine, which some critics predicted would limit its reach and may have initially prevented theater owners from wanting to take a chance on the film, turned out to be a large part of its appeal.

Of the 828 reviews on IMDb, forty-four specifically mentioned the focus on wine as an asset to the film, calling it a "must see for wine lovers" or professing to have enjoyed "the wine aficionado stuff" more than expected.[44] The film also served a pedagogical purpose for audiences eager to learn more about wine. Another fifty described the film as informative or educational about wine and California's wine country. For example, one IMDb user wrote, "Where else will you learn how to taste wine properly, have a good laugh and relate to two of the most charming losers ever seen on film—all at once?"[45] For viewers who didn't find the losers so charming, the wine education might have been the only appeal. An IMDb user who slammed the film with a 1 out of 10 rating on IMDb said that they "did enjoy the shots of wine country and wine tasting 101."[46] User-submitted reviews were more likely than reviews by film critics to describe the movie as slow or boring, but not usually because of the wine. For example, a comment on *Metacritic* called *Sideways* "boring as hell," but then said that "the movie's wine theme is actually pretty interesting and not only do you learn about all sorts of wine, each character takes on their own type."[47]

Anecdotal reports from the wine industry suggest that the film affected how people across the country thought about wine, particularly the key varietals Merlot and Pinot Noir. In February 2005, two months after the film's wide theatrical release, a Seattle sommelier said that at least two or three customers a night were specifically mentioning the movie when ordering, some asking for Pinot Noir and others suddenly sheepish about ordering Merlot, prefacing their order with, "'I'm know I'm not supposed to.'"[48] In March 2005, an Ohio wine store owner was quoted in the *Sunday Times* of London as saying, "People have been coming in and asking for the *Sideways* grape, even if they don't quite remember its name."[49] The idea that liking Merlot was something of a faux pas was echoed by Virginia Madsen, the actress who plays Maya, in an interview with the *New York Times*. She described a recent visit to the Los Angeles restaurant Pastis: "'They fooled me,' she said. 'They brought out this wine and we were like, this is really good, thinking it was the pinot as usual.' It turned out to be a Merlot: horrors. 'If you saw it on a menu, you'd throw it across a room. It was a merlot from Malibu.'"[50] Even some people who may not have seen the film seem to have caught wind that Merlot was now uncool. Without mentioning *Sideways*, Katie Couric said on *The Today Show* that she had heard she wasn't supposed to drink Merlot.[51]

The film actually had a measurable impact on Merlot and Pinot Noir sales, too. In January 2005, ACNeilsen reported that the number of households buying Merlot was down 2 percent compared to the same twelve-week period in the previous year, although overall wine sales were up.[52] The change in Pinot Noir sales, a much smaller segment of the wine market, was even more striking: from October 24, 2004 (two days after the film's limited opening weekend) to July 2, 2005, grocery-store sales of Pinot Noir jumped 18 percent.[53] According to a team of economists at Sonoma State University led by Steven Cuellar, the change in the demand and price for Merlot and Pinot Noir was both statistically significant and lasting.[54] Cuellar and his colleagues examined the sales volume and price of Merlot, Pinot Noir, and several other varietals that didn't feature as prominently in *Sideways*—Cabernet and Syrah—before and after the film's release. Based on annual scan data from U.S. retail chains from 1999 through 2008, they found that until 2004, the sales growth rate of all the varietals increased at a similar rate. After 2004, Merlot sales slowed or even declined slightly while Pinot Noir sales increased precipitously. The change in the demand for those varietals varied significantly from the unfeatured control varietals, with Merlot growing less and Pinot Noir growing more than Cabernet and Syrah. They also found a decrease in the price of Merlot and an increase in the price of Pinot Noir that was consistent with those changes in demand. They concluded that "all the results are consistent with the theory that *Sideways* had a negative impact on the consumption of Merlot, while increasing the consumption of Pinot Noir."[55]

One possible explanation for the "*Sideways* effect" is that *Sideways* enlightened its audience. The effect might be evidence the film taught Americans to be better wine drinkers by educating them about how bad Merlot is—or at least a lot of the Merlot they were previously buying—and encouraging them to seek out Pinot Noir, a varietal many likely had not tried before. However, wine critics generally claim the opposite. They argue that Pinot Noir, or at least most of the Pinot Noir Americans have been buying since *Sideways* came out, isn't inherently better than Merlot and is getting worse as the demand rises. In short, they claim that the film didn't educate people, it duped them into thinking that a varietal is a reliable heuristic for quality. According to *New York Times* wine critic Eric Asimov, the "*Sideways* effect" was exaggerated by the media and the only thing the film had really done was flood the market with a "growing sea of bad Pinot Noir."[56]

If much of the Pinot Noir flooding the market was bad according to the experts, why were people buying it anyway? Especially given that Pinot Noir is associated with the snobbish, loser Miles before his burger redemption, it might seem counterintuitive that sales rose after *Sideways*. If the film really endorses the less pretentious approach to wine represented by Maya, Pinot

Noir shouldn't have fared any better in the wake of the film's release than Merlot or Cabernet. Indeed, that's exactly how Mary Baker, the owner of a small winery in Santa Barbara County, explains the fact that her Merlot sales weren't hurt: "No one wants to be the 'geeky Miles.' Miles apparently hated Merlot, and that made people curious about Merlot because no one wants to come across as such a navel-lint-gazing wine snob."[57] However, Baker's experience clearly conflicts with broader market trends and probably only applies to the tiny niche market occupied by people who might purchase a bottle of $40 Merlot from a self-described "microwinery" in Paso Robles, California—the same tiny group who might have understood why Miles drinking a bottle of Cheval Blanc was ironic. The passionate defense of Merlot many wine critics mounted also seems symptomatic of some people's desire to appear less pretentious by embracing the maligned grape or by seeking to distinguish themselves from the masses they saw as having been duped into drinking the flood of bad Pinot Noir.

Ironically, then, the only people who made the connection between Pinot Noir and snobbery were those who might themselves be considered wine snobs. Since most wine consumers didn't know that the Cheval Blanc was a Merlot blend, Miles's invective against Merlot had more popular resonance than its role in his redemption did. Instead of creating an association between Pinot Noir and snobbery, Sideways turned Pinot Noir into sign of sophistication and Merlot into a sign of mediocrity that the casual consumer believed someone with real wine know-how would shun. Furthermore, the film reassured viewers that cultivating the kind of good taste represented by Pinot Noir wouldn't necessarily make them snobs because Miles is ultimately redeemed. What audiences seemingly learned about wine from Sideways is that it's possible to have good taste without being pretentious about it. The film not only offered viewers an education about wine, it also gave them permission to put that education into practice and use their new culinary capital without feeling like navel-gazing snobs.

FROM GOURMETS TO FOODIES: EVADING THE TAINT OF ELITISM

Despite the popularity of texts such as the Grey Poupon commercials and the film Sideways that negotiate with the threat of snobbery, anxieties about elitism in the realm of food are tenacious. The idea that food is too mundane to constitute a real form of snobbery or that real connoisseurs enjoy both luxury and lowbrow foods might sometimes give aspirational eating a partial or provisional pass. However, the popular discourse about food reveals persistent suspicions that people are using food to gain status and judging other people when they fail to adhere to the new standards of culinary propriety. One example of how this suspicion attaches to trends with high-status culinary capital is the adoption of and then backlash against the term foodie. The term was initially embraced in part because it was perceived as free from some of the elitist baggage of older words such as "gourmet" or "epicure." Over time, though, "foodie" has also become associated with elitism, and now even people who identify food as one of their primary interests tend to chafe against it.

Like the word "snob," "foodie" reflects something of the historical moment in which it was adopted. The term appears to have been independently coined by both *New York Magazine* restaurant critic Gael Greene in 1980 and the anonymous author of a letter to the features editor of *Harpers & Queen* magazine in 1981, in the early years of the contemporary food revolution.[58] Its origins also offer a hint of the conflicts to come, as it was unclear from the start whether the term was meant to be an insult or badge of pride.

Greene first used the word "foodie" in an article titled "What's Nouvelle? La Cuisine Bourgeoisie" to describe the clients at a hip restaurant in Paris whose chef, Dominique Nahmias, was at that time the only woman to rate three red toques from Gault Millau (the equivalent of three Michelin stars). She described the scene: "She offers crayfish with white feet or red . . . three ways, tends stove in high heels, slips into the small Art Deco dining room of *Restaurant d'Olympe*—a funeral parlor of shiny black walls and red velvet—to graze cheeks with her devotees, serious foodies, and, from then on, *tout Paris*, the men as flashily beautiful as their beautiful women."[59] Although the word "bourgeoisie" now refers to primarily to the elite, the title of Greene's piece refers to a movement that was initially seen as less sophisticated and snooty than *nouvelle cuisine*, in a throwback to the older European class system in which the bourgeois were middle-class upstarts encroaching on the aristocracy. *La cuisine bourgeoisie* was how Parisians referred to the growing interest in regional, seasonal, traditional cooking, "like Grandma used to make," that was spearheaded by chefs such as Nahmias, who had no formal culinary institute training. Later in the article, Greene described her as "a housewife friends pestered to turn professional."[60] Thus, the term "foodies" not only differentiates the "serious" eaters from the "flashily beautiful" people who show up after ten just to be seen but also refers specifically to restaurant-goers who had embraced a style that departed from the French culinary establishment. In other words, these were connoisseurs, not snobs.

Around the same time that Greene was sampling *la cuisine bourgeoisie* in the shifting culinary landscape of Paris in 1980, Ann Barr, editor of the features section of *Harpers & Queen* in England, noticed that "the food world was shifting on its tectonic plates, and that perfectly sane people had suddenly become obsessed with every aspect of food."[61] When she invited readers to send her their thoughts on the phenomenon, several seized on the opportunity to criticize regular *Harpers* contributor Paul Levy. One of the letters referred to him as a greedy, gluttonous, lip-smacking "king foodie." Levy and Barr were immediately excited about the word—Levy later described it as "a cocktail stick applied to a raw nerve"—and they began working on a satirical field guide to foodies similar in style and tone to *The Yuppie Handbook*. *The Foodie Handbook* was published in 1984 and is sometimes erroneously credited with inventing the term, although Barr and Levy credit both Greene and the anonymous letter writer.[62] They defined the foodie as "a person who is very very very interested in food" and "consider[s] food to be an art, on a level with painting or drama."[63] By 1989 the word had achieved sufficient cultural currency to appear in the *Oxford English Dictionary*.

Although the anonymous letter writer clearly used the word in a pejorative sense, Levy claimed that it lost its negative edge as soon as more people began to use it: "What started as a term of mockery shifted ground, as writers found that 'foodie' had a certain utility, describing people who, because of age, sex, income, and social class, simply did not fit into the category 'gourmet,' which we insisted had become 'a rude word.'"[64] To call someone a gourmet had effectively become a way of calling them a snob and carried connotations of being old, male, rich, white, and conservative—none of which necessarily applied to the people participating in the new food movement. The term caught on because people needed a new word to differentiate between the stodgy, staid reputation the term "gourmet" had acquired during the decades that fancy food spent on the cultural margins and the predominantly young, liberal population leading the new food movement.

In 2007, reflecting on the term "foodie" nearly three decades after he helped popularize it, Levy claimed, "It long ago stopped being (if it ever really was) a term of abuse. But is it a compliment about your knowledge of food or the sensitivity of your palate? Or is it simply a value-neutral description, like civil servant, football fan, or stamp-collector?"[65] Four of the eighteen people who commented on the article agreed that foodie was a value-neutral description and that food was a hobby like any other. Several expressed surprise that it had ever had negative connotations. Two said they considered it to be a term of praise, and seven claimed to use the term to describe themselves with pride. Three suggested that the debate about whether or not it's a good thing to care about what you eat could only be had in the United Kingdom and the United States because everyone in countries like Italy and France care about good food. But four disagreed vehemently with Levy. As one wrote, "No, it is still a term of abuse in my book. Please stop using the F-word. I am not a

f**die!" Another rejected it specifically because of the association with elitism, claiming the term had become "over-used" and "seems to have simply
replaced 'gourmet[,]' which sort of negates the function assigned to it in the
article, particularly regarding class assumptions."⁶⁶

For some people, "foodie" may always have been objectionable. Particularly
for those who might be gratified by the idea that their knowledge about food
or the sensitivity of their palates might set them apart, "foodie" represents the
increasing popularity of food that renders their knowledge and talents less
remarkable. It's roughly the equivalent of the cool independent band whose
obscurity offers cultural cachet becoming a pop sensation. The diminutive
ending also lacks a certain dignity. Nonetheless, the term must have been
innocuous enough before 2010 for the growing resistance against it to be
identified as a backlash. In a 2010 *Houston Press* blog post titled "Has the
'Foodie' Backlash Begun?" Katharine Shilcutt reported that all the urban
alternative weekly papers in the Village Voice Media chain had recently been
instructed to avoid the term because of its negative connotations. As she
explained, "A foodie is no longer someone who appreciates and enjoys food.
A foodie is now someone who takes food to extremes: Tweeting every course
of every meal, obsessively discussing *Top Chef Masters* and *Hell's Kitchen*
episodes on Internet forums, forcing the entire group to wait as they take
pictures of every dish that hits the table and rushing to upload them to Flickr
or Twitpic, grilling wait staff to find out the exact provenance of every ingredient in a dish and revering chefs as if they were Lennon and McCartney."⁶⁷
The way she distinguishes between an older, more acceptable foodie-ism and
the newer, excessive form is similar to the distinction between Maya and
pre-burger Miles: the former appreciate and enjoy food, the latter obsess
and engage in status competition. Again, it seems that liking food is fine,
and maybe even a good thing; it's only when that appreciation seems like a
performance designed to impress other people and win culinary cred that it
becomes unappealing.

The backlash against the term began around the same time as a wave of
criticism and parody of the food revolution as a whole began. In November
2010, *South Park* devoted an entire episode to skewering foodies and celebrity
chefs, portraying an obsessive interest in the Food Network and ingredients
such as crème fraîche as symptoms of the pathetic sexual frustration of middle-aged suburbanites.⁶⁸ In December 2010, a *Chicago Tribune* food writer
declared "Foodie Fatigue," complaining about the "obsession-of-the-moment
and one-upmanship and casual snobbery and need to proclaim *this* the most
authentic Mexican cooking in Chicago or *that* a poor excuse for a Korean
taco."⁶⁹ In October 2012, a *Huffington Post* headline claimed "Foodie Backlash

Has Come Swiftly, Was Inevitable." The article reported similar frustrations with the focus on food as a "high-end lifestyle" and the obsession with all things local, free-range, gluten-free, whole grain, artisanal, or from some obscure geographical provenance.[70] The heart of all of these critiques was that the interest in food—which the articles explicitly claim was itself a good thing—became objectionable when it was used to jockey for social position, either as a performance of one's own privilege or grounds for disparaging judgments about other people. For example, Amanda Hesser, a food writer and co-founder of the online community *Food52*, told the *Chicago Tribune* columnist: "Knowing a lot about food culture is a good thing. . . . I'm pro-food experts. I'm just not so sure I want to have dinner with them or have them judge me on the coffee I drink."[71]

The word "foodie" might initially have seemed like a less elitist way to refer to food enthusiasts, but the creeping association with practices that are also seen as excessively status conscious eventually caught up with it. Some people still self-identify as foodies, but they often qualify their embrace of the term by specifically differentiating themselves from food snobs. Those attempts at differentiation frequently take the form of gesturing toward the lowbrow foods they love or are at least willing to eat, particularly American cheese and fast food. Like Miles, they engage in culinary slumming to prove that their appreciation for food is egalitarian.

Both those who reject the term and those who embrace it with qualifications acknowledge that at least some of the recent interest in food is more about social positioning than the food itself, but no one is willing to admit that their own interest in food could be that crass or venal. It's always other foodies who are snobs. Other foodies are blind to the privilege involved in the search for rare artisanal cheeses and authentic tamales. Other foodies scorn the contentment of the undiscerning masses with drip coffee, Budweiser, and well-done corn-fed steaks served with instant mashed potatoes and MSG-laced gravy. Everyone admits that snobbery happens, and many even seem to think that snobbery based on food has become rampant. But no one wants to admit that their own tastes and judgments have been affected. At the same time, almost everyone yearns on some level to have their tastes sanctioned by their peers. As long as food forms the basis for social judgments, anxieties about elitism will haunt the culinary practices that carry elite cultural capital.

DISTINCTION, DEMOCRACY, AND DENIAL

In *Foodies*, Johnston and Baumann argue that the term "foodie" embod-ies the tension between democracy and distinction, which they call "two competing poles in the gourmet foodscape."[72] They describe them as follows: "A democratic pole that eschews elite cultural standards and valorizes the cultural products of 'everyday' non-elite people, and a pole of distinction that continues to valorize standards that are rare, economically inaccessible, and representing significant amounts of cultural capital."[73] According to Johnston and Baumann, the tension between these poles explains why foodies are different from the gourmets and epicures of the past, whom they characterize as concerned only (or at least primarily) with distinction. Like their predecessors, foodies sometimes seek out the most expensive meals and ingredients. But unlike their predecessors, they also idealize authenticity and exoticism, which Johnston and Baumann argue are "reasonable and potentially egalitarian criteria—not snobbish."[74]

However, Johnston and Baumann imply that that their own affection for cheap, lowbrow ethnic and comfort foods is what distances them from foodies. They open the book by confessing: "We have never really considered ourselves foodies" and describe their tastes as bridging the high-low divide:

> At the same time we unconsciously acquire knowledge of high-end food establishments, we frequent an extensive roster of lower-cost eateries selling foods like bi bim bop, lamb roti, and barbeque brisket. We try to eat organic and local foods as frequently as possible, but we are not above tucking into food court french fries, or using the occasional bribe of fast food to obtain compliance from our children. We enjoy the challenge of learning how to make unusual food items at home, like marshmallows or osso buco, but we frequently fall back on homey staples like rice pudding, roast chicken, and cinnamon buns.[75]

After a few more examples, they ask, rhetorically, "Given our obvious affection for foods of various varieties and genres—high and low, fast and slow—doesn't this make us foodies?" Instead of answering that question, they launch into a discussion of how the term has always been contested.

Their own reluctance to identify with the term and their use of lowbrow examples to counter the idea that they are full participants in the "gourmet foodscape" is a bit like Miles's burger, designed to ensure that the reader knows they are on the side of inclusion, not exclusion. And yet the implication that their wide-ranging tastes might disqualify them from being foodies suggests that the relationship between democracy and distinction isn't a simple antagonism between two poles or a continuum with good, inclusive foodies on one end and bad, snobbish ones on the other. Instead, like most participants in the food revolution (whether they self-identify as foodies or not) Johnston and Baumann use their appreciation of lowbrow food to legitimize their own elite culinary capital. It's precisely this kind of gesture toward inclusiveness that serves to obscure how food reproduces class hierarchies.[76]

Although some self-identified foodies claim they are willing to eat anything, the purported omnivorousness of the gourmet foodscape is frequently overstated. The range of foods that can carry high-value cultural capital may have diversified—although even that notion is suspect, as aspirational eaters in the late nineteenth and early twentieth centuries were interested in far more than just French food—but diverse isn't necessarily the same thing as inclusive. Most of the new markers of culinary superiority create complementary forms of culinary inferiority. Today's food enthusiast might buy and enjoy both French's yellow mustard and Dijon, but that doesn't mean that the products communicate the same thing. Furthermore, it is the ability to buy and enjoy both, not in spite of the differences in their status but because of them, that distinguishes the contemporary culinary sophisticate.

The performance of omnivorousness itself communicates privilege. Many advocates of the cultural democratization theory, including the sociologists Richard Peterson and Roger Kern and John Seabrook of the New Yorker, have noted the declining significance of the distinctions between highbrow and lowbrow that emerged in Britain and the United States in the eighteenth century. According to Peterson and Kern, surveys of preferences regarding music and art reveal a "qualitative shift in the basis for marking status—from snobbish exclusion to omnivorous appropriation. . . . Highbrows are more omnivorous than others and they have become increasingly omnivorous over time."[77] Similarly, in the book Nobrow, Seabrook argues that contemporary cultural capital, especially that marked by coolness and buzz, is less exclusive, more democratic, and more meritocratic than the taste hierarchies that set the elite of his parents' generation apart, such as their appreciation for classical music, wine, and opera. At the same time, many previously low-status genres have been elevated.

However, the rise of rock 'n' roll and $400 denim jeans doesn't necessarily represent a decline in "snobbish exclusion." Instead, as elites have become more omnivorous, their ability to discuss and display an appreciation for a wide range of music, literature, art, fashion, and food has become their most distinguishing characteristic. Comfort and familiarity with both expensive and difficult-to-acquire things and cheaper, more accessible ones are the sine qua non of today's savvy cultural consumer. In fact, many of the things most likely to attract buzz are cheap but still difficult to acquire—the trendy local band that still plays in dive bars for a $10 cover or the hole-in-the-wall diner that serves the most authentic tacos or bibimbap. Although not exclusive in the same way that tickets to the opera or Bordeaux wines are, the cheap but obscure also requires knowledge and access that most people don't have. Demonstrating one's omnivorousness might seem like a way of resisting traditional cultural hierarchies, but it also indicates a breadth of cultural experience that is now associated with the elite in part because it requires exceptional resources. Lacking an appreciation for diverse cultural forms and showing a parochial attachment exclusively to the lowbrow, the mainstream, and the mass produced is now the mark of the lower class.

People can be quite sensitive to shifts in the social significance of their consumption choices, as is demonstrated by the spread of rumors that Merlot was uncool or déclassé, and yet not be fully aware of how that shapes their tastes. One's own tastes typically feel natural and inevitable. Some foods simply seem appealing, others appalling. Even if we know on some level that those preferences reflect aspects of our upbringing and personality, including class status, the way our mouth waters or our nose wrinkles at the smell of a hot dog seems to come from somewhere deep within, not the kind of posturing associated with snobbery. The visceral experience of taste mediates the use of food as a performance of identity and a basis for social judgments. Most people who purchased Pinot Noir or avoided Merlot after *Sideways* probably wouldn't say that their purchases were motivated by concerns about how others perceived them. Some of them might not have even remembered why Pinot Noir suddenly seemed appealing and Merlot less so. Instead, if they were conscious of any motivation, it was probably taste. They were trying to purchase wine they would enjoy. However, as the research on taste perception suggests, their enjoyment is inevitably shaped by indicators of status, such as price.

Buzz and rumor also influence how classy or trashy a particular food seems to someone, how much they like it, and what they think about other people who eat it or don't eat it. When mediated by taste preferences and expectations, the performance of status through food may not feel like snobbery. Even behaviors that seem baldly performative, like affecting a French accent when you order a croissant or swirling a tasting pour of wine might not be done in a conscious effort to impress. But the tastes and habits you acquire from your upbringing, your peers, mass media texts, and the whole swirl of culture you live and breathe communicates things about your class status whether you want it to or not.

Johnston and Baumann claim that the foodies they interviewed are largely unaware of class and they describe the popular discourse of the gourmet foodscape as "classless." However, the concern about elitism that animates the Rolls-Royce commercials, *Sideways*, and debates about the word "foodie" constitutes a tacit, widespread acknowledgment of the fundamental inter-relatedness of class and social judgments about taste. If people don't readily admit that class performance is one of their own motivations for eating "better," that doesn't mean they're unaware of class. The many strategies present in mass media texts for negotiating with the threat of food snobbery suggest that people are acutely aware of class. What they may be unaware of, or simply reluctant to acknowledge, is how their own preferences and even their most immediate, visceral reactions to food are affected by the social rewards for having good taste.

CHAPTER 6

Feeling Good about
Where You Shop

SACRIFICE, PLEASURE, AND VIRTUE

We're talking about health, we're talking about the planet, we're talking about the people who are supporting the land. . . . Make a sacrifice on the cellphone or the third pair of Nike shoes.

—*Alice Waters*

It's not just the consumption of gourmet foods that must be defended against the threat of being perceived as snobbery. The consumption of natural foods and the burgeoning movement in support of local and sustainable food have also been dogged by accusations of elitism. The idea that those trends are the exclusive preoccupation of the bourgeois is responsible for their inclusion in parodies such as the blog *Stuff White People Like*, which is as much a send-up of the professional middle class as it is of whiteness. The full list ranks farmers' markets #5, organic food #6, vegan/vegetarianism #32, and Whole Foods and grocery co-ops #48.[1] The entry on natural foods markets claims that they have taken the place of churches and cathedrals as the most important buildings in white communities and then extends the metaphor: "Many white people consider shopping at Whole Foods to be a religious experience, allowing them feel good about their consumption. The use of paper bags, biodegradable packaging, and the numerous pamphlets outlining the company's [policies] on hormones, genetically modified food and energy savings. This is in spite of the fact that Whole Foods is a profit driven-publicly traded corporation that has wisely discovered that making white people feel good about buying stuff is outrageously profitable."[2] The blog's author adopts an air of exposing a hidden ideology with the "in spite of the fact" phrasing, but the big secret is made explicit by the company slogan that used to be printed on those conscience-gratifying paper bags: "Feel Good About Where You Shop."

Many of the leaders and participants in the sustainable food movement have been eager to explain why making consumption choices you can feel good about isn't elitist. Their primary tactic is an admission that yes, the movement has been primarily restricted to a relatively wealthy few so far due to the higher cost of good food, but that doesn't make it elitist. What's really elitist is the system that prevents most people from having enough money, time, or knowledge to eat good food all the time. For example, in an interview about the backlash against the food movement, Michael Pollan told Ian Brown of the *Toronto Globe and Mail* that "to damn a political and social movement because the people who started it are well-to-do seems to me not all that damning. . . . The reason that good food is more expensive than cheap food is part of the issue we're trying to confront."[3] In other words, the movement may reflect the inequalities that shape our society, but it's really working to fight those inequalities. If anything, it is populist, not elitist. It's notable that he contrasts good food with cheap food, not with bad food, as if it is impossible that food could be simultaneously cheap and good.

Alice Waters has also admitted that good food tends to be more expensive, but insists that it's worth it. The quote in the epigraph suggests that anyone can make what she has called "more satisfying choices" if they're willing to revise their spending priorities. Her example of three pairs of Nike shoes carries some ugly racial connotations, invoking the 1990s-era stereotypes about the profligate poor that focused on urban black communities. Additionally, it may be a little arrogant, if not elitist, to presume that she knows what uses of their money people will find more satisfying. But like Pollan, her aim seems to be reframing the sustainable food movement as populist. It's just that the masses she's concerned about are farmers, not consumers.[4]

Others dispute the assumption that good food is more expensive. In 2007, Nina Planck, the founder of London Farmers' Markets and New York City's Greenmarket, told *Plenty* magazine: "Organic food and whole food—what I call traditional food—is frugal. Buy a whole chicken. It serves four people twice—the second time as soup. Buy fresh, local produce in season and canned wild Alaskan salmon. . . . In the center aisles are processed, nutrient-poor, high-profit-margin foods. That's what will eat up your budget."[5] By inverting the typical representation of junk food as cheap and organic food as expensive, she also recasts the food movement as populist, but in the sense of already being accessible to anyone who knows how to shop strategically. Similarly, the former *New York Times* columnist Mark Bittman often wrote that cooking what he called "real food" like a roast chicken and salad for dinner is cheaper—and, according to him, tastier—than feeding a whole family at McDonald's.[6] They imply that if there's a barrier to broader participation in the movement, it is ignorance, not cost.

A third tactic the food movement's defenders use is dismissing the debate about elitism as misleading and possibly even invented for that purpose by public relations firms and lobbyists who represent the food industry and big agricultural interests. In a 2011 *Washington Post* opinion column titled "Why Being a Foodie Isn't Elitist," Eric Schlosser claimed that the elitist epithet was a "form of misdirection" used by groups such as the American Farm Bureau Federation to "evade serious debate" about things such as the current agricultural subsidy system, which disproportionately benefits the wealthiest 10 percent of farmers. According to Schlosser, "It gets the elitism charge precisely backward. America's current system of food production—overly centralized and industrialized, overly controlled by a handful of companies, overly reliant on monocultures, pesticides, chemical fertilizers, chemical additives, genetically modified organisms, factory farms, government subsidies and fossil fuels—is profoundly undemocratic. It is one more sign of how the few now rule the many."[7] Schlosser's portrayal of the food movement as a democratic revolt against a real elite represented by the food industry echoes Waters's pleas to support small farmers and Planck's reference to "high-profit-margin" processed foods in the center aisles. However, he sidesteps the question of cost entirely. Similarly, in the documentary *Food, Inc.*, Joel Salatin responds to questions about whether it's elitist to expect people to pay what he charges for free-range eggs and meat by declaring the issue "specious."[8]

These attempts to defuse or dismiss the debate about elitism haven't been very successful. It persists like a proverbial thorn in the food movement's side. One need not be a shill for the food industry to wonder whether a movement based largely on buying more expensive things (the essence of "voting with your fork") can ever include enough people to inspire the desired reforms.[9] Like the pervasive threat of food snobbery in representations of gourmet culture, debates about elitism in the movement for sustainable, ethical food may reveal the troubled conscience of people whose participation in those trends is based at least in part on the performance of a desired class identity. However, the issue of elitism in the food movement is more concerned with issues of price and access than with taste.

This chapter discusses three examples of conflicts over price and access in the food movement and how they relate to the moralization of bodily pleasures such as sex and food. The first, Slow Food USA's 2011 "$5 Challenge," demonstrates that far from being a deterrent, the expense and inconvenience that makes eating better a "challenge" are central to its appeal for the elite.

Michel Foucault's history of sexuality helps explain how this kind of willful denial of pleasure comes to seem virtuous, particularly to the elite. The second example is the popular discourse about the extent to which price governs food choice, which reveals widespread assumptions about the relationship between cost and morality and suggests that the high price of better food is a less important deterrent for most people than food activists often assume. Instead, it seems that price may actually matter most to wealthier consumers who use it as a signal of more virtuous consumption. Lastly, a *Salon* article about hipsters using food stamps that went viral in 2010 demonstrates how the consumption of these foods is policed. The vitriol inspired by the idea that people might use public assistance to buy organic foods seems to undermine the notion that natural foods are really more ethical. What may be justified as a virtuous sacrifice for the elite suddenly looks like luxury spending when it comes to the poor.

A Generational Shift in Slow Food's Negotiations with Elitism

Slow Food USA is a branch of the organization started in Italy in 1986 by a writer named Carlo Petrini that now has chapters called convivia in over 150 countries. According to the statement of philosophy on the Slow Food movement's website, their guiding principles are "that the food we eat should taste good; that it should be produced in a clean way that does not harm the environment, animal welfare, or our health; and that food producers should receive fair compensation for their work."[10] Slow Food convivia pursue those ideals by hosting potluck dinners, starting gardens, organizing classes and workshops, and holding fund-raisers to support local food producers. National branches, which exist in seven countries, provide leadership and resources for the convivia and coordinate larger projects such as the Ark of Taste, an archive of historical recipes and heirloom seeds and livestock breeds.

The relationship between the U.S. branch and some of its convivia began to sour after the national organization began to move in a more political direction under the leadership of a new president named Josh Viertel. Viertel is a Harvard graduate who started an organic farm in Massachusetts and then co-founded the Yale Sustainable Food Project. When asked by *Forbes* magazine to name "The World's 7 Most Powerful Foodies" in 2011, Michael Pollan ranked Viertel number three, saying, "He has moved the American wing of this international organization front and center on questions of access and policy, while continuing to celebrate the cultural and biological diversity of our food traditions."[11] However, many members chafed against the new direction, and tensions reached an apex after the launch of the $5 Challenge campaign.

The campaign began in August 2011 when Slow Food USA issued a press release inviting people to "take the $5 Challenge" by pledging "to share a fresh, healthy meal that costs less than $5—because slow food shouldn't have to cost more than fast food."[12] According to Viertel, the goal of the campaign was to prove that sustainable agriculture can be populist rather than elitist.[13] The national chapter may also have been trying to diversify its membership, which skews white, wealthy, and old. Instead, the primary result of the campaign was to drive away longtime supporters, including Alice Waters, fellow chef-activist Deborah Madison, and the founders of several convivia. Four months into the campaign, the popular food website *Chow* (formerly *Chowhound*) published an article titled "Cheap Drama at Slow Food" that claimed that the leadership had been "embroiled in a bitter squabble . . . stoked by angry emails, hurt feelings, accusations."[14] The primary complaint of Waters and other critics of the campaign was that promoting cheap food represented a betrayal of what they saw as the central mission of the Slow Food movement, which is prioritizing food quality, environmental sustainability, and fair wages for farmers over cheaper prices. According to Poppy Tooker, who founded the New Orleans convivium in 1999 before the national chapter even existed, "We had spent all these years trying to make sure that the farmers were championed and other food producers were paid a fair wage for what they brought to our tables. The $5 Challenge put a hole in that."[15]

One interpretation of the controversy might be that the $5 Challenge exposed the elitism of the sustainable food movement or at least of the old guard represented by people such as Waters and Tooker. For example, in response to a *SFoodie* blog post titled "Slow Food's $5 Challenge Made Alice Waters Cry," one reader noted, "Alice may have been a hippie at some point in her life, or hippie-ish, but she's catered to a moneyed crowd for a long time now," and another affirmed, "Amen. What's the point of organic, sustainable food if the only people who can eat them are the rich. . . . The idea that slow food dinners cost 80 dollars seems like some sort of terrible way of keeping the rich healthier and the poor fatter."[16] By seeking to demonstrate how Slow Food could be accessible to people on a limited food budget, supporters of the $5 Challenge saw it as a noble attempt by the young, idealistic Slow Food USA leaders to drag older members such as Waters into the new dawn of food justice.

However, despite the campaign's populist rhetoric, its architects were seemingly ill acquainted with the resources actually available to the poor. According to the 2011 USDA Food Plans used to calculate supplemental nutrition allowances, the minimum cost of a nutritionally adequate diet averages $1.43 per meal, and approximately 15 percent of the American population cannot even afford that without assistance. In fact, even the most liberal 2011 USDA food budget allotted only $3.73 per meal for food cooked at home.[17] Far from being a modest budget, $5 per person would be extravagant for most Americans. Nonetheless, the Tumblr offering "Tips and Tricks" for preparing a meal for the $5 Challenge is clearly pitched at an audience for whom the challenge is not affording a $5 meal but rather restricting themselves from spending more. For example, one entry from December explaining how to "add zing to your $5 Challenge meal" suggests: "Try incorporating that honey-ginger hot sauce you bought on an island vacation ages ago, or sprinkle the fleur de sel you paid a premium for but doesn't cost a dime toward your $5 meal!"[18]

Even if it had gained more traction, the $5 Challenge may have done more to exclude working-class Americans than prove that they too can partake of the pleasures of the Slow Food table. There's a paradox inherent in the campaign: on the one hand, it says that eating slow food on a limited budget is hard, it's a "challenge," but on the other hand, it suggests that anyone can do it. The challenge doesn't do anything to combat or even call attention to the structural barriers that might make it harder for some people to get fresh, local foods at prices they can afford. Instead, it individualizes the problem by implying that anyone willing to put in the time and effort can live up to the slow food ideal. The campaign reinforces the widespread assumption that anyone can make healthy, delicious, conscientious food choices if they want to and are willing to make some sacrifices. Participation in the sustainable food movement is portrayed as dependent only on an individual's willingness to forgo the easy pleasures of fast, cheap food.[19] Ultimately, that means anyone who doesn't eat the foods currently deemed morally superior has only herself to blame.

Thus, both the critics and supporters of the $5 Challenge ultimately reinforce the exclusivity of Slow Food. However, the conflict between them reflects different strategies for negotiating with the threat of elitism. The Waters camp justifies the decision to spend more money on food by framing it as the only way of ensuring that farmers are paid fair wages. For them, slow food cannot be both cheap and fair. The younger Slow Food USA staffers, on the other hand, want to make sustainable food accessible to everyone in a tautological fashion by proving that it already is. For them, slow food must be cheap to be fair.

The $5 Challenge ultimately garnered a grand total of 5,572 pledges, a relatively modest rate of participation for an organization that claimed at the time to have over 25,000 members and 250,000 supporters.[20] Aside from limited coverage in specialty media such as *Chow* and on blogs such as *Grist* and *Culinate*, the campaign attracted virtually no media attention. The few articles it did inspire were largely devoted to the controversy it caused.[21] Additionally, *Chow* reported that declining donations after the start of the $5 Challenge forced the national office to lay off five staffers from the Brooklyn headquarters and temporarily suspend work on projects such as the Ark of Taste.[22] When Viertel resigned in June 2012, it was widely attributed to the organization's inability to agree on its approach to issues such as accessibility.[23]

Although the controversy certainly didn't help the campaign any, its failure was probably preordained by the mistaken assumption that the primary barrier to more people participating in the food revolution is access. The $5 Challenge seemed to suggest that if only people knew that they could afford to eat fresh, locally grown, healthy food, surely they would. It's the "if you build it, they will come" philosophy of sustainable food. But what if the desires and anxieties that drive some people to buy organic, local, fair-trade, grass-fed, sustainable, local, healthy food aren't universal? The $5 Challenge exemplifies the ethos of sacrifice that unifies the sometimes disparate and competing camps in the food revolution and the limitations on the movement's broader appeal. Although the $5 Challenge was designed to show that anyone can participate in the Slow Food movement, its failure instead demonstrates that most Americans don't want to.

The Right Use of Pleasure

The moralization of self-indulgence as bad and self-denial as good dates back at least to the fourth century b.c. The ancient Greeks are sometimes thought of today as hedonistic pagans who exulted in sodomy and bacchanalia. However, in *The History of Sexuality* Michel Foucault argued that their moral philosophers actually glorified self-restraint and moderation, particularly in the realms of food, drink, and sex.[24] In the *Nicomachean Ethics*, Aristotle distinguished between those realms, which he categorizes as pleasures of the body, and the loftier pleasures of the mind, such as delighting in paintings, music, or the theater. According to Aristotle, only the former are prone to overuse and self-indulgence (*akolasia*): "There is pleasure that is liable to *akolasia* only where there is touch and contact: contact with the mouth, the tongue, and the throat (for the pleasures of food and drink) or contact with other parts of the body (for the pleasure of sex)."[25] Furthermore, Aristotle argues, excesses of food, drink, and sex can be ruinous, whereas virtually no one destroys their lives by overindulging in the love of art or even sensual pleasures where there is no contact, such as the smell of a rose.

Although the inclination to excess made bodily pleasures seem morally dangerous to the ancient Greeks, it also made them useful for honing the self-discipline that was thought to demonstrate maturity. Foucault argued that for the Greeks, food and sex were equivalent moral challenges: "Foods, wines, and relations with women and boys constituted analogous ethical material; they brought forces into play that were natural, but that always tended to be excessive; and they all raised the same question: how could one, how must one 'make use' (*chrēsthai*) of this dynamics of pleasures, desires, and acts? A question of right use. As Aristotle expresses it, 'all men enjoy in some way or another both savoury foods and sexual intercourse, but not all men do so as they ought [*ouch' hōs dei*].'"[26] It was through the moderation (*sōphrōsyne*) of bodily pleasures such as food and sex that free male subjects were expected to demonstrate the capacity for self-governance that justified their participation in the governance of the body politic. Women and slaves, on the other hand, were seen as inherently incapable of controlling their bodily appetites, a belief that was used to justify their exclusion from the assembly.

Early Christians also saw elective asceticism as the path of virtue and developed elaborate regimens of dietary and sexual restrictions that were supposed to foster both physical and spiritual health. According to Foucault, Christian theology of the first and second centuries was characterized by a pervasive "mistrust of the pleasures" and an "emphasis on the consequences of their abuse for the body and the soul."[27] Eating right was no simple matter. The fifth-century writer Orisbasius devoted four entire books to "the qualities, disadvantages, dangers, and virtues of the different possible foods and to the conditions in which one should and should not consume them."[28] It was also in this period that a particular concern about the moral consequences of overeating began to emerge. Preachers began delivering sermons against gluttony that declared it a cardinal vice because it represented an impious worship of the senses and weakened the defenses against drunkenness and lustful acts.[29] With the development of monastic orders, chastity and fasting became important ways of demonstrating faith and virtue. Religious women especially used the renunciation of food as a way of renouncing sensual pleasures and identifying with the suffering of Christ on the cross. According to Carol Walker Bynum, "Self-starvation, the deliberate and extreme renunciation of food and drink, seemed to medieval people the most basic asceticism, requiring the kind of courage and holy foolishness that marked the saints."[30]

Foucault said that by the Victorian era, sex had surpassed food in importance and had become a distinct realm of moral concern. He argued that the modern idea of sexuality emerged, along with the idea that people could be defined by their sexual behaviors, at the end of the eighteenth century.[31]

Instead of merely proscribing particular acts such as sodomy as immoral, the new language of sexuality created whole new identities—the homosexual, for example. Foucault's argument about the emergence of sexuality at the turn of the nineteenth century was the inverse of his argument about the ancient Greeks. Whereas the Greeks were supposed to be hedonists, Victorians are now widely seen as being prudish and sexually repressed. Contrary to the idea that the Victorian bourgeoisie made sexuality taboo and silenced sexual discourse, Foucault argued that there was a "veritable discursive explosion" about sex in the late eighteenth and nineteenth centuries.[32] New institutions such as school dormitories, hospitals, and the bourgeois family established elaborate rules about sexual propriety and technologies of surveillance and control focused on uncovering sexual "perversions," such as adultery, masturbation, and homosexuality. However, he ignored parallel developments in the moralization of food.

As many other scholars have pointed out, there was also a great proliferation of discourses and institutions aimed at regulating the consumption of food and drink during the eighteenth and nineteenth centuries, at the same time that Foucault claimed the new technology of sex emerged. It was in the late eighteenth century that alcoholic beverages and other drugs first were distinguished from food and their use was subjected to the discourses of medicine, psychiatry, and social reform.[33] Dr. Benjamin Rush, one of the signers of the Declaration of Independence, is credited with founding both the American temperance movement and American psychiatry. Uniting these interests, he criticized alcohol and tobacco use in medical and moral terms, pioneering the language of addiction to talk about excess drinking and the prescription of psychiatric treatment for alcoholics.[34] The "moral thermometer" in his *Inquiry into the Effects of Ardent Spirits upon the Human Body and Mind* exemplifies attitudes toward the bodily pleasures of food and drink that developed around the turn of the nineteenth century (see Fig. 17). He equated the temperate liquids at the top of the chart, for example water and buttermilk, with "health, wealth, serenity of mind, reputation, long life and happiness" while he associated anything more alcoholic than "weak punch" with long lists of vices, diseases, and punishments. Moderation and morality were also central to Rush's writings on tobacco. He claimed that statistical observation had revealed that tobacco activated and worsened diseases of the nerves and complained of its filthiness and associations with idleness and rudeness. Noting that tobacco was alleged to provide relief from "intemperance in eating," he wrote, "Would it not be much better to obviate the alleged necessity of using Tobacco by always eating a moderate meal?"[35]

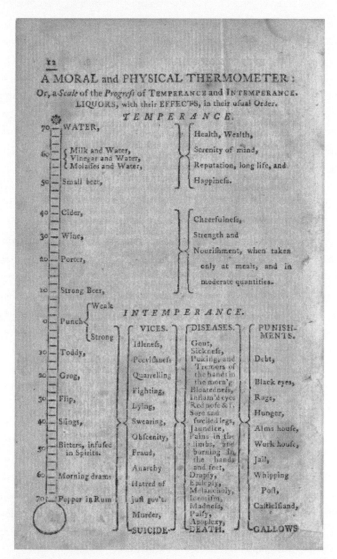

Figure 17. Medicalizing and moralizing food and drink at the turn of the nineteenth century. Source: Benjamin Rush, *An Inquiry into the Effects of Spirituous Liquors on the Human Body and the Mind*, 1790. The Library Company of Philadelphia. Reproduced by permission.

As with the new discourse of sexuality, the bourgeois family was one of the main sites where the new rules about eating and drinking moderately were cultivated. These new rules helped produce a distinctive bourgeois body and way of being. Middle-class families in the eighteenth and nineteenth centuries came to see mealtime as a crucial opportunity for training children in manners, conversation, and taste. Children began to dine with their parents instead of with nannies and servants as it gradually became seen as improper to delegate that important opportunity for moral education to the working-class help.[36] Much of the practical advice written for Victorian mothers focused on proper care of the adolescent female body in particular, including how their daughters should eat and exercise to cultivate the correct social identity and moral character. The recommendations often reflect the connections people continued to make between food and sex; meat and spicy foods were thought to stimulate and signal sexual desire, so their consumption was seen as unsuitable for proper ladies.[37] The combination of smothering maternal concern about eating and the elevation of restraint and physical delicacy has been blamed for the emergence of anorexia nervosa among middle-class girls in the Victorian era.[38]

Advocates of temperance, domestic scientists, and health crusaders such as Sylvester Graham and John Harvey Kellogg were also deeply invested in moralizing both food and sex, which were often seen as mutually constitutive appetites. Both Graham and Kellogg claimed that vegetarianism could reduce sexual urges. Writing in 1891, Kellogg claimed that "the science of physiology teaches that our very thoughts are born of what we eat. A man that lives on pork, fine-flour bread, rich pies and cakes, and condiments, drinks tea and coffee, and uses tobacco, might as well try to fly as to be chaste in thought."[39] His use of "the science of physiology" to justify his claims echoes the medicalization of sexual perversions Foucault described. Kellogg wrote that the person seeking to avoid unchaste thoughts should "*discard all stimulating food. Under this head must be included spices, pepper, ginger, mustard, cinnamon, cloves, essences, all condiments, pickles, etc., together with flesh food in any but moderate quantities.*"[40] This advice was later echoed by home economists who expressed concerns that the diets of immigrant populations—particularly their use of spices and pickled foods—would indulge the appetite for stimulating foods and fail to teach children self-denial.[41]

The growing interest in and anxiety about food in the three decades since Foucault's death has led some people to question whether it has now surpassed sex in importance. For example, a 2009 article in *Policy Review* by Mary Eberstadt asked "Is Food the New Sex?" She argued that shifts in the availability of food and sex and the potentially negative consequences of both have created an unprecedented situation in the West where most adults can have practically all the food and sex they want. But, she noted, instead of pursuing both with equal ardor or developing more limitations on both kinds of appetites, a surprising paradox has emerged between "mindful eating" and "mindless sex."[42]

To illustrate, she asked readers to imagine the sexual and culinary mores of a hypothetical woman named Betty, who was thirty years old in 1958, and her thirty-year old granddaughter, Jennifer. Betty eats mostly industrially processed foods and has no strong feelings about what other people choose to eat; Jennifer shops for organic groceries and has very strong convictions about how other people should eat. Conversely, Jennifer is laissez-faire about sexual ethics, believing that things such as abortion, STDs, and homosexuality are fundamentally personal matters; Betty believes all three are clearly wrong and the world would be a better place if people restricted themselves to procreative sex within heterosexual marriage.

Ultimately, Eberstadt concluded that the increased vigilance about food reflects the triumph of modern nutritional science: "Decades of recent research have taught us that diet has more potent effects than Betty and her friends understood, and can be bad for you or good for you in ways not enumerated before."[43] The mystery, she said, is why people haven't come to the same conclusion about sex, given that some researchers have also concluded that monogamous, married couples live longer, happier lives. She claimed that the apparent struggles of our ethical and legal institutions to cope with a sexually laissez-faire society suggest that "mindless sex" is as unnatural and bad for us as "mindless eating." Eberstadt suggested that we need restrictions on our carnal appetites. Noting that people are often furtive about their indulgence in both junk food and junk sex, Eberstadt wrote, "It is hard to avoid the conclusion that the rules being drawn around food receive some force from the fact that people are uncomfortable with how far the sexual revolution has gone—and not knowing what to do about it, they turn for increasing consolation to mining morality out of what they eat."[44]

Both Foucault and Eberstadt essentially portrayed morality as a zero-sum game in which any increase in the ethical concern about food is countered by an equal and opposite decrease in the concern about sex, and vice versa. Foucault seems to have assumed that sex surpassed food as a subject of moral concern; the first volume of the *History of Sexuality* was an inquiry into the moral significance of sexual acts. However, his focus on sex may have prevented him from exploring the proliferation of discourse about the moral significance of food in the Victorian era. The moralization of food, eating, and the body in the Victorian era challenges the notion that sex was ever a separate or more important form of bodily pleasure. Eberstadt proposed that the new morality of food is an attempt to compensate for the conspicuous lack of morals in a sexual revolution gone too far. However, her theory relies on the mistaken assumption that the older generation represented by Betty is amoral about food and the younger generation represented by Jennifer is amoral about sex.

Betty's eating does not conform to the new norms of the food revolution, but that does not make her eating normless. The norms that govern Betty's eating are apparent in Eberstadt's description:

> Betty's freezer is filled with meat every four months by a visiting company that specializes in volume, and on most nights she thaws a piece of this and accompanies it with food from one or two jars. If there is anything fresh on the plate, it is likely a potato. Interestingly, and rudimentary to our contemporary eyes though it may be, Betty's food is served with what for us would appear to be high ceremony, i.e., at a set table with family members present. . . . The going slogan she learned as a child is about cleaning your plate, and not doing so is still considered bad form.[45]

The meat-and-potatoes meals that Betty prepares for her family are not mindless; they merely reflect a very different set of beliefs about what it means to eat right. During World War II, the idea that fighting men deserve meat served to solidify its place in the iconographic dinner.[46] The mandate to serve meat at every meal has also been connected to a belief in the basic human right of domination—over nature, over animals, over resources, and over other people.[47] The expectations that a woman should do the cooking and the family should eat together reinforce traditional gender norms and the high social importance placed on the nuclear family.[48] Buying frozen meat from a company that specializes in volume and insisting upon cleaning your plate reflect frugality and an anti-waste ethic. Betty's meals place a high value on tradition and consistency and on the idea that people should be grateful for what they're given instead of expecting meals that cater to their own individual appetites or preferences.

Similarly, Jennifer's embrace of casual sex, homosexuality, and reproductive choice is not necessarily "mindless." Jennifer is part of a generation that has come to reject, in many cases quite thoughtfully and deliberately, the practice of shaming or discriminating against people for having sex outside heterosexual marriage.[49] This ethos of tolerance and egalitarianism reflects a shift in the nature of sexual mores, not an abandonment of them.

Eberstadt underestimated the importance of food as a site of moral concern for the older generation, the continued moral importance of sex for the younger generation, and the persisting connection between food and sex. The use of food and sex are still as central to the constitution of ethical selfhood as ever. Instead of a realignment in the importance of the bodily pleasures, the contrast Eberstadt described between the Betty generation and the Jennifer generation reflects the sea change in norms associated with the sexual revolution and the food revolution. Her characterization of Betty's eating habits and

Jennifer's attitude toward sex as "mindless" compared to their counterparts who follow elaborate rules about what not to eat and what kind of sex should be avoided conflates virtue with elective restriction.

Here's how Eberstadt describes Jennifer's diet: "Wavering in and out of vegetarianism, Jennifer is adamantly opposed to eating red meat or endangered fish. She is also opposed to industrialized breeding, genetically enhanced fruits and vegetables, and to pesticides and other artificial agents. She tries to minimize her dairy intake and cooks tofu as much as possible. She also buys 'organic' in the belief that it is better both for her and for the animals raised in that way, even though the products are markedly more expensive than those from the local grocery store."[50] Jennifer's choices are based mostly on avoidance. She never eats red meat or endangered fish and sometimes avoids animal products entirely. She avoids pesticides, industrially produced meat, and GMOs. She also sacrifices money and time, buying "markedly more expensive" food somewhere other than the local grocery store. Compared to Betty's attachment to mid-century cuisine, which is limited more by habit than elective self-deprivation, the new way of eating that Jennifer practices seems more ethical because it follows in the long tradition of demonstrating virtue through sacrifice.

WHO TAKES PLEASURE IN RENOUNCING PLEASURE?

Foucault's explanation of how the discourse of sexuality developed and spread in the Victorian era can help explain the class dynamics of an ethics based on sacrifice. Foucault argued that the deployment of Victorian sexuality did not happen evenly across social classes. It proceeded in roughly three stages. First, it was embraced by the bourgeoisie, whose initial concern was the creation of a distinctive body: "The primary concern was not the repression of the sex of the classes to be exploited, but rather the body, vigor, longevity, progeniture, and descent of the classes that 'ruled.' This was the purpose for which the deployment of sexuality was first established. . . . The class which in the eighteenth century became hegemonic . . . provided itself with a body to be cared for, protected, cultivated, and preserved from many dangers and contacts, to be isolated from others so that it would retain its differential value."[51] Next, sexuality and the institutions devoted to the creation of this "'class' body," especially the nuclear family, were extended, often forcibly, to the working class.[52] As Foucault put it, the family "came to be regarded, sometime around the eighteen-thirties, as an indispensable instrument of political control and economic regulation for the subjugation of the urban proletariat."[53] Finally, at the end of the nineteenth century, sexuality was institutionalized, especially in the professional discourses of medicine and law. Foucault identified this as "the moment when the deployment of 'sexuality,' elaborated in its more complex and intense forms, by and for the privileged classes, spread throughout the entire social body."[54]

The first decades of the food revolution parallel the first stage of the emergence of sexuality. When the professional middle class first started trying to eat better again in the 1980s, giving up cheap processed foods and choosing more expensive natural ones, the focus was on cultivating their own healthy bodies and distinctive practices. The second stage, the moralization of the poor, began around 2000 with the declaration of a war on obesity and the sudden proliferation of attempts to get the lower classes to eat according to values that had become dominant in the professional middle class.[55] A quintessential example is the Los Angeles City Council's unanimous vote in July 2008 to place a moratorium on the opening of fast-food restaurants in low-income areas, citing above-average rates of obesity in neighborhoods such as South Central.[56] Also in 2008, New York City mayor Michael Bloomberg signed Local Law 9, which established 1,000 permits for "green carts" that can only sell whole fruits or vegetables and are restricted to low-income neighborhoods.[57] The limited success of these initiatives may also hint at a resistance on the part of the poor similar to the Victorian proletariat's refusal to accept the sexuality "foisted on them for the purpose of subjugation."[58] A study released by the RAND Corporation in 2015 revealed that the ban on new fast-food restaurants had "failed to reduce fast food consumption or reduce obesity rates in the targeted neighborhoods."[59] Of the 350 green cart permits made available in Brooklyn, only 84 were in use two years later, and many vendors complained that they couldn't compete with local stores that have a better selection of produce and are allowed to sell other foods.[60]

Part of the problem with policies designed to make the poor eat better is a flaw in the assumption that they're eating badly to begin with. Although the idea that poor people have inadequate access to fresh produce has become an article of faith for many advocates of food system reform, the actual research on access is mixed. The USDA defines a "food desert" as any census tract with a poverty rate of 20 percent or greater or a median family income at or below 80 percent of the area median in which at least 500 residents or 33 percent of the population lives a mile or more from the closest supermarket or full-service grocery store. According to the USDA Economic Research Service (ERS), that means that 23.5 million Americans live in food deserts, more than half of whom are low income.[61] However, smaller or more specialized corner stores, street vendors, and farmers' markets are all excluded from the ERS analysis, which might be why several studies have found that living in an area defined as a "food desert" doesn't have any relationship to most measures of food insecurity or obesity.[62] As Roland Sturm, one of the authors of a study on food access in California, told the *New York Times*, "Within a couple miles of almost any urban neighborhood you can get basically any type of food. . . . Maybe we should call it a food swamp rather than a desert."[63] It's not clear that lack of access is a real barrier to eating better for most people.

Policies that aim to get poor people to eat more local and organic food may also fail because the ethics of sacrifice that make those things seem superior is simply less appealing to people who can't reap the symbolic rewards as easily. There have always been some people who take pleasure in elective asceticism. Even as gluttony became a cardinal vice for early modern Christians, manuals of confession warned of stricter penalties for the "arrogance of excessive fasting" than for overindulgence in food.[64] Some people can even enjoy feeling hungry, especially if they believe that overeating is sinful and fasting is virtuous. The people most likely to find rewards in giving up easy, convenient pleasures are those who feel guilty or conflicted about their potential overconsumption, a trait that is especially characteristic of today's liberal elite.

Price, Pleasure, and Sacrifice

If there is a single principle that unifies the four ideals of the food revolution, it may be that eating better requires giving up some of the conveniences and pleasures of the conventional industrial food system. Eating gourmet, healthy, natural, or cosmopolitan requires eschewing readily available, widely beloved foods such as Lucky Charms, Kraft macaroni and cheese, McDonald's French fries, and Coca-Cola. The pursuit of more complex, hard-to-get, and less widely appreciated foods can be enjoyable too. The Slow Food movement describes its philosophy as a reclamation of the pleasures of cooking from scratch, celebrating foods in season, and lingering over a meal. However, those pleasures all depend on giving things up—money, time, and the effort required to cultivate a taste for foods that aren't as easy to love such as kale and Brussels sprouts. For the food revolution faithful, those sacrifices aren't a deterrent but are actually central to their perception that eating gourmet, healthy, natural, and authentic food is morally superior.

Most of the de facto spokespeople for the food revolution claim that the main reason anyone is still eating Doritos and McDonald's despite the increasing availability of locally grown produce and grass-fed steaks is that the former are so much cheaper. Michael Pollan and Alice Waters do not deny that the food they want people to eat is more expensive. They depend on that fact to explain why their preferred foodways have not yet become universal.

If fresh, local, organic food were both inherently better and cheaper, presumably everyone would be eating it. According to Pollan, the problem is that the higher cost of real food tips the scales in favor of junk.

In a 2007 *New York Times* article, Pollan described a study by obesity researcher Adam Drewnowski that found that if he had a single dollar to spend at the typical American supermarket, he could buy 1,200 calories of cookies or potato chips and 875 calories of soda but only 250 calories of carrots or 170 calories of orange juice. "This perverse state of affairs is not, as you might think, the inevitable result of the free market," Pollan explained, but instead is the result of archaic agricultural policies that subsidize cheap, unhealthy sources of calories such as commodity corn and the meat from corn-fed animals.[65] Since the Agricultural Act of 2014, also known as the Farm Bill, does almost nothing to subsidize the production of fresh fruit and vegetables, the "rules of the food game in America are organized in such a way that if you are eating on a budget, the most rational economic strategy is to eat badly—and get fat."[66] Pollan accordingly welcomes increasing prices for fuel and food on the grounds that they "level the playing field for sustainable food that doesn't rely on fossil fuels."[67]

However, there's some evidence that for the average consumer, price isn't the primary factor governing food choices. According to the ERS, from 1980 to 2006—precisely the period when fast food supposedly overtook the U.S. diet and made Americans the fattest people on the planet—food prices declined across the board. The price of apples, dry beans, carrots, and celery went down right along with the price of cookies, ice cream, and potato chips.[68] According to the ERS, "The price of a healthy diet has not changed relative to an unhealthy diet."[69] Writing about the ERS data in *The Atlantic*, James McWilliams concluded that "evidently, consumers have chosen to take advantage of the declining prices for the cookies rather than the apples, thereby undermining the claim that we choose cheap unhealthy food because it's cheap. As it turns out, we also choose it because we appear to like it better than cheap healthy food."[70]

Ironically, there's some evidence to suggest that people who are truly eating on a budget eat less of the most prevalent symbol of cheap, unhealthy food. Fast-food consumption rises with income, peaking in American households with incomes in the $60,000–$80,000 range.[71] A 2013 Gallup poll found that Americans in households with an annual income over $50,000 were more likely to say they eat fast food on a weekly basis than those in lower-income groups.[72] In 2015, the Centers for Disease Control found that low-income children get fewer calories from fast food than wealthier ones.[73] According to a survey commissioned by the anti-hunger organization Share Our Strength, low-income Americans eat meals prepared at home more often than richer people, and the poorer they are, the more likely they are to cook meals from scratch.[74] Whenever one of these studies makes its way to the media, the findings are treated as if they're new and surprising. Just like Nina Planck and Mark Bittman, who wrote about the frugality of home-cooked meals as if it were their original discovery, many people simply assume that the poor eat more fast food and cook less than the middle class. However, budget constraints effectively prevent them from doing so. For people with a little more money, food choices appear to be influenced less by price than by taste. And for slow food advocates who avoid cheap food to the extent that devising a $5 meal is a challenge, reverse price signaling is at work.

For the people who seek out expensive foods because they believe them to be superior, the sacrifices involved aren't a deterrent. They are central to their perception that their foodways are better. At one point in an interview with the *Wall Street Journal*, Pollan adopted an uncharacteristically neutral position about local food, saying "To eat well takes a little bit more time and effort and money. But so does reading well; so does watching television well. Doing anything with attention to quality takes effort. It's either rewarding to you or it's not. It happens to be very rewarding to me. But I understand people who can't be bothered, and they're going to eat with less care."[75] However, his entire *oeuvre* is devoted to making the case that eating well will be rewarding for everyone, in part because it takes more time and effort and money.

For example, in the first chapter of *The Omnivore's Dilemma*, Pollan wrote, "Though much has been done to obscure this simple fact, how and what we eat determines to a great extent the use we make of the world—and what is to become of it. To eat with a fuller consciousness of all that is at stake might sound like a burden, but in practice few things in life can afford quite as much satisfaction. By comparison, the pleasures of eating industrially, which is to say eating in ignorance, are fleeting."[76] Far from portraying junk food as an innocent predilection, such as watching reality television instead of *The Wire*, Pollan insists that eating determines nothing less than "the use we make of the world." Rather than just one of many satisfying things one could devote time and effort to if one were so inclined, he portrays eating well as the only rational choice that someone who is fully conscious of the true costs and benefits could make. Eating industrial food is not the rational, morally neutral choice of someone who simply "can't be bothered"; it is "eating in ignorance."[77]

Pollan suggests that the reason eating well matters is because of its effect on the world, but when the environmental or political effects of local or organic foods are questioned, many proponents of local food admit to being more interested in the symbolic act of voting against the industrial food system than in the real effects of their consumption choices. As discussed in chapter 3, many people argue that food that is produced in optimal conditions and is shipped to markets in large quantities tends to have a smaller per-unit carbon footprint than locally grown produce. The latter often requires greater inputs and takes a less efficient path to market. James McWilliams made that case in his 2009 book *Just Food: Where Locavores Get It Wrong and How We Can Truly Eat Responsibly*. A review by Stephanie Ogburn posted on the environmental site *Grist* claims that McWilliams misses the point, suggesting that the real effects of local farming aren't locavores' main concern: "Consumers, when faced with a system they don't support, are voting with their dollars for the only alternatives they can find-local food at the farmers market and organic products at the store. . . . The locavores I know don't view shopping consciously as a solution; they view it as a protest."[78] Ogburn suggests that what matters to locavores is not the actual carbon footprint of their purchases so much as the symbolic protest buying any alternative to industrially produced food represents. The real goal is distinguishing themselves from the norm.

Other reviewers claim that there are unquantifiable benefits to eating locally or that what matters is that locavores are making an effort. A review of McWilliams's book by Kelly Trueman published on *AlterNet* claims that McWilliams fails to take into account the meaningful relationships and communities forged at farmers' markets, "as opposed to the soulless commerce of the supermarket."[79] Trueman says that McWilliams "dwells obsessively on food miles, presumably because he couldn't acknowledge these benefits without undermining his own arguments." But it's unclear how the possibility of forging meaningful relationships over carrot purchases might undermine McWilliams's argument that the carbon footprint of local food is larger instead of smaller, as its proponents often claim. A review in the *Christian Science Monitor* by Rebekah Denn praises McWilliams for "digging beneath slogans and oversimplifications" but concludes that "it seems counterproductive to simultaneously belittle those already trying to make the best choices they can."[80] None of the reviewers seem willing to consider the possibility that the best choice for anyone attempting to eat sustainably might be choosing the products of conventional, industrial agriculture over local and organic alternatives.

In contrast, all three reviewers praise the one part of McWilliams's book that supports the practice of virtuous self-denial: his claim that the best way to eat more sustainably is to eat less meat. Trueman calls the book a "cogent critique of America's unsustainable addiction to meat . . . buried in a mound of manure about 100-mile diet diehards," and Ogburn says, "The most sensible recommendation McWilliams makes is that if we want to lesson agriculture's impact on natural systems, we need to eat less meat."[81] Denn says of his argument about local foods, "the bad taste this leaves in the mouth is a shame, because his other points are important and pressingly relevant," singling out his support for "reducing the amount of meat in our diets, a move that he correctly notes would save more energy than many well-meaning dietary habits."[82] None of them explain why his evidence in that chapter is any more convincing than in the chapters that dispute the purported advantages of local and organic production or the evils of GMOs. This response to *Just Food* suggests that the desire to eat better isn't driven by evidence about what's really healthier, more sustainable, more humane, or even better tasting. If eating better resembles what most people are already doing, such as buying conventional goods, it simply isn't appealing. Like these reviewers, many advocates of sustainable food are only interested in eating better to the extent that it is difficult and thus distinguishes them from people unable or unwilling to put in the effort.

Proof of the nutritional or ecological superiority of particular foods is ultimately unnecessary to make people feel like choosing them is virtuous. Making a special trip to the farmers' market during the few hours per week it's open seems like a virtuous act in part because it's so much less convenient than shopping at a grocery store that's open all the time. Turning a box of locally grown produce you probably never ate as a child into edible meals must be better—morally, if not nutritionally—than microwaving a Lean Cuisine frozen meal. Spending more money on something with an organic or hormone-free label must be better, because otherwise why would it cost more? People don't even need to know what it's better for to reap the psychic rewards of self-denial. Things that are difficult, inconvenient, or require sacrifices do offer the satisfaction Pollan promises—at least for some people. The idea that sustainable food is virtuous relies on the deeply held belief that it is good to work hard and resist immediate pleasures. Participants in the food revolution might claim that what makes the hard work and sacrifice worthwhile is some long-term goal or objective good, but their resistance to attempts to evaluate the real impact of their choices suggests otherwise.

FOODIE SHAME: POLICING THE BOUNDARIES OF TASTE AND CLASS

A belief in the superiority of sustainable food is not limited to the elite. But while people such as Pollan and Waters might applaud Americans of all income levels who are spending more money on the kinds of food they see as superior, many others express discomfort or even anger at the prospect of poor people buying foods presumed to be more expensive. An article in *Salon* by Jennifer Bleyer titled "Hipsters on Food Stamps" published in March 2010 and the responses to it exemplify the deep popular ambivalence about the relationship between the virtues of eating better and the class status of the eaters. Bleyer describes what she claims is a growing trend prompted by the recession that started with the 2008 financial collapse: "Faced with lingering unemployment, 20- and 30-somethings with college degrees and foodie standards are shaking off old taboos about who should get government assistance and discovering that government benefits can indeed be used for just about anything edible, including wild-caught fish, organic asparagus, and triple-crème cheese."[83]

Her evidence is mostly anecdotal. Bleyer went grocery shopping with an unemployed art school graduate and a part-time blogger with a degree from the University of Chicago who both lived in a neighborhood in northwest Baltimore that she describes as artsy. They had recently qualified for $150–$200 a month in supplemental nutrition assistance, which they used to buy ingredients intended for a Thai curry, including lemongrass, coconut milk, a Chinese gourd, and clementine juice. She also interviewed an AmeriCorps volunteer from Brooklyn who supplemented his stipend with food stamps (an approved part of the standard AmeriCorps compensation) and who said that some of his artist friends who lived in Williamsburg had also qualified for nutritional assistance in recent months. The final piece of evidence was a quote from a cashier at Rainbow Grocery in San Francisco who said that more young people seemed to be buying organic food with electronic benefits transfer cards.[84]

Bleyer admitted that the scope of the phenomenon was difficult to measure, perhaps because "hipster" is not a demographic recognized by human resource administrators. The food policy experts she interviewed insisted that most food stamps went to "traditional recipients: the working poor, the elderly and single parents on welfare." She also mentioned the dramatic increase in unemployment for people from the ages of twenty to thirty-four. From 2006 to 2009, the rate doubled for that age group and increased 176 percent for those with a bachelor's degree or higher, leading Bleyer to grudgingly admit that "young urbanites with a taste for ciabatta may legitimately be among the new poor." But then she quoted Parke Wilde, a food economist from Tufts, who implied that recent college graduates who find themselves in poverty usually have other sources of financial support: "There are many 20-somethings from educated families who go through a period of unemployment and live very frugally, maybe even technically in poverty, who now qualify." Particularly when the word "educated" is applied to entire families rather than individuals, it is more likely a proxy for economic privilege than a comment on their credentials, and the idea that they deserve any kind of support is further undermined by the description of their poverty as technical and temporary.[85]

The article generated over 450 comments in three days, at which point *Salon* closed the thread. A substantial majority were from angry readers demanding that government assistance be limited to more deserving people and that it restrict beneficiaries from purchasing expensive luxuries. Many readers sought to distinguish the hipsters described in the article from the legitimate, deserving poor, using their high-class tastes as the primary evidence that people such as the ones described in the article don't belong on government assistance. For example, one reader wrote, "Instead of asking the government to assist your educated palate, maybe you should be asking Mom & Dad. Obviously they gave you the taste of the good life to begin with, which you feel you are entitled to."[86] This commenter conflated the concept of education with elite status. She believes that college degrees and high-class tastes make people seem like illegitimate recipients of government support, regardless of their income.

Readers were more likely to question the authenticity of the hipsters' poverty than whether the food Bleyer described them buying was really all that expensive. For example, one wrote, "If a single person qualifies and can shop in this fashion every day then 1) he or she is receiving too much money or 2) they run out of money mid-month and are getting money from somewhere else (mommy and daddy) and should not qualify to begin with."[87] The insistence that buying Chinese squash and coconut milk must reflect wealthy social origins and involve excessive expenditures attests to the perceived exclusivity of the practices associated with foodies. Many other readers suggested that the people described in the article must be unemployed by choice, either on the assumption that they could get jobs in the service industry if they were willing to do them or should have known better than to major in something as impractical as art. One writes, "I have a hard time believing these people truly exhausted their options. Did they look into fast-food, janitorial services, retail? . . . No—these type of jobs don't fit into their self-image as an artist or whatever. So, even though this is a situation of their own making—they're expecting the government to subsidize their lifestyle. And its all being paid for by people who actually bite the bullet and work at jobs they don't necessarily love. . . . Yeah, it's pretty appalling."[88]

The reproachful comments fit a long tradition of stigmatizing poor people as lazy, self-indulgent, and the authors of their own impoverishment. Welfare recipients have been especially demonized since the 1960s, when the stereo-typical recipient of direct government aid shifted from a temporarily unem-ployed white man to a chronically dependent black woman.[89] However, unlike the indolent "welfare queen" that Ronald Reagan made one of his primary talking points in his presidential campaign, who supposedly used fifty differ-ent Social Security numbers to collect thousands of dollars a month in public handouts and drove a Cadillac,[90] all that was required to cast aspersion on Bleyer's hipsters was their college degrees and their ability to cook a Thai curry. They weren't accused of greed or gaming the system, at least not as overtly as the supposed welfare queen was. The suggestion that they might have unde-clared income support from their families is a considerably more modest accusation than the ones Reagan used to cast aspersions on the public safety net. Instead, the primary focus of readers' anger is that they appear to be, as the above reader says, living "the good life" and therefore cannot be truly poor.

The primary debate in the comments concerned whether the eating habits Bleyer described are as self-indulgent as the article makes them out to be or might be seen as defensible or even virtuous. In the opening paragraph, Bleyer said that hipsters from Baltimore "sauntered through a small ethnic market stocked with Japanese eggplant, mint chutney, and fresh turmeric," the sauntering implying an insouciance about the use of food stamps that was later belied by their shame and defensiveness about relying on public assistance. Most of the examples of upscale food the article mentioned were conjured from Bleyer's imagination and were not foods she actually observed the hipsters buying or eating, aside from the curry ingredients. She used these icons of high-status food culture to undermine her theoretical defense of their practices: "Food stamp-using foodies might be applauded for dem-onstrating that one can, indeed, eat healthy and make delicious home-cooked meals on a tight budget. And while they might be questioned for viewing pre-mium ingredients as a necessity, it could also be argued that they're eating the best and most conscious way they know how. . . . Is it wrong to believe there should be a local, free-range chicken in every Le Creuset pot?"[91] They might be applauded, she hedged, but then she conflated the act of making healthy and delicious meals on a tight budget with "viewing premium ingredients as a necessity," an implicit contradiction. She further undermined the idea that what they were doing might be seen as thrifty in her updated version of the proverbial chicken in every pot that invoked the ethical labels that typically demand a price premium and an expensive French brand of cookware.[92] The subtitle of the article also explicitly invited readers to criticize the trend that she may have entirely invented: "They're young, they're broke, and they pay for organic salmon with government subsidies. Got a problem with that?"[93]

Most people who responded to the article issued a resounding "Yes," but a small minority defended the hipsters for eating real food instead of junk despite their limited means. These readers frequently suggest that the hipsters' superior food choices would prevent them from getting fat and taxing the health care system. For example, one wrote, "With rising obesity epidemics and other diet-related health issues so prevalent in our culture, why would we want these folks to spend their meager allotment on highly processed foods laden with fats and high fructose corn syrup instead of organic carrots, salmon, and other healthy items?"[94] Others echoed Planck and Bittman in reframing real food as frugal rather than extravagant, like the reader who applauded the hipsters for what they presumed is a principled refusal to support agribusiness: "I have absolutely no problem with this. Its about time that food stamps were used for real food. Its about time that people really learn how to cook food as opposed to buying canned food and chips. Its the best way to stretch a buck. It also take's away from the profits of the like's of Nabisco, Nestle, and all of those other corporations who don't care and produce corn and other products."[95] This reader specifically portrayed buying healthy food as thrifty and buying junk food as profligate. Others also claimed the hipsters deserve praise for not relying on the conventional cheap foods that are presumed to be fattening. For example, one wrote, "They are buying in the store rather than McDonalds and they are getting something healthy and cooking it well. . . . Should people suffer and eat junk food that makes them diabetic (which those working for would have to pay for)?"[96] Another suggested that a better solution than preventing poor hipsters from buying good food would be to make them teach traditional food stamp recipients how to shop and cook.[97] The hipsters' defenders romanticized them as resourceful and frugal, people who were admirably resisting the excesses of the industrial food system.

Like Reagan's myth of the welfare queen, Bleyer's hipsters on food stamps became a repository for anxieties about class, which is likely why the article generated such an enthusiastic response. A typical 2010 article in *Salon* received around a dozen comments. More than three dozen was unusual, and 400 or more was truly rare. What almost all the readers who left comments had in common, whether they described the hipsters as admirable or loathsome, was that their judgment was based primarily on whether they interpreted the food choices described in the article as a renunciation of pleasure or as indulgence in it. Many of the people who argued that food stamps should be further restricted specifically complained about the possibility that people could use government assistance for pleasure. For example, one wrote: "ONLY BASIC foods should be OK for food stamps. No chips, no cakes, no artisanal breads, nothing fancy[. . . .] There are millions of non food stamp people buying beans and rice to save money while food stamp folks can buy fun food? No, that isn't right. If I could wave a magic wand . . . I would say ONLY basic vegetables, fruits, beans, and grains are OK for foods stamps. Not much else."[98] The reason this reader gave for ruling out both junk foods and fancy foods wasn't how expensive the foods were but whether or not they were "fun."[99]

Perhaps part of the reason so many readers were so quick to mandate thrift and asceticism for the hipsters was that it enabled them to displace the shame they might feel about their own culinary indulgences. Just as the image of obese poor people buying frozen pizzas and soda is often a useful target for fat shame, the hipsters with their heaping bowls of government-subsidized squash curry drew attention and derision because they are the ideal repository for *foodie* shame. This eagerness to heap criticism on other people for their imagined culinary pretension and profligacy may be one of the reasons so many readers credulously accepted such a thin story and even made their own unfounded assumptions about what kinds of things hipsters on food stamps might buy. Bleyer either discovered or invented the perfect whipping boy for anyone who fears that their own participation in contemporary food trends might be a kind of self-indulgence instead of self-denial.

The negative response to the idea of poor people eating fancy food also points to the difficulty of acquiring high-status cultural capital if you're not already part of the group associated with it. The very same foods that have been popularly portrayed as virtuous when they're seen as being primarily consumed by the elite, such as sustainably caught fish or free-range chicken, suddenly become luxurious excesses when people imagine the poor eating them. When the moral valence of a cultural sign shifts with the class status of the person, that's probably a good sign that the normative judgment associated with the sign has more to do with social hierarchies than with any real moral logic. Just as the threat of being perceived as a snob helps police nonwealthy people who might aspire to pass as gourmets, the vitriol inspired by the idea of poor people participating in the food movement helps keep those trends exclusive enough to continue serving as a source of symbolic distinction.

It remains to be seen whether the ideals of the food revolution will ever be as fully institutionalized as Victorian era sexual ethics eventually became. However, the longer history of the ideals of the food revolution suggests that the relationship between class formation and the moralization of bodily pleasures may not always follow precisely the pattern Foucault described. According to Foucault, the "new regime of discourses" represented by sexuality took shape along with the bourgeoisie during the development of capitalism. He saw subsequent shifts within that discourse, like the apparent decline in sexual prohibitions in the twentieth century, as less significant than the continued use of sexuality to define and control people's behavior.[100] When it comes to food, shifts within the dominant discourse seem to correspond to changes in the relative influence of different social classes. Self-denial may always maintain its association with virtue, but the popular desire to embrace that virtue has varied significantly over time. The two periods when the culinary ethics of self-denial, represented especially by the ideals of thinness and purity, have been culturally dominant are the ones characterized by extremes of inequality and the stagnation in the growth of middle-class wealth.

The relationship between social class and ideology doesn't mean that the people who embrace the ideology are simply performing what they've perceived to be a desirable status marker without truly believing in it. Foucault's argument wasn't that the bourgeoisie specifically set out to craft a distinctive body by merely pretending to believe that certain sexual behaviors were perverse because they thought it would differentiate them from the masses. Nor did he argue that they applied their new technologies of surveillance and new categories of perversion to the poor with a deliberate agenda of class subjugation in mind. Instead, as the bourgeoisie became the dominant social class in the capitalist system, the members of that class were inclined to see themselves as different from the proletariat and therefore they embraced beliefs and practices that affirmed that difference. They really believed that masturbation and homosexuality were wrong, but their investment in those beliefs was dependent on their status.

Similarly, the professional middle class didn't embrace the ideals of the food revolution because they realized that they would require sacrifices they knew the poor wouldn't be able to afford or want to make. There's no reason to doubt that the vast majority of people who eat local food because it's supposedly better for the environment really believe that that is true. What the $5 Challenge, foodies' reverse price sensitivity, and concern about whether hipsters on food stamps are being profligate or thrifty all suggest is that the investment in beliefs such as the superiority of local food is based on a deeper association between sacrifice and virtue that is more likely to appeal to the liberal elite. As long as eating better is constructed as being dependent on hard work and sacrifice, the food revolution is going to continue to appeal primarily to that elite. Meanwhile, their efforts to get other people to join them will continue to seem elitist to everyone who isn't as eager to demonstrate their moral superiority through their consumption choices.

Conclusion

CONFRONTING THE SOFT
BIGOTRY OF TASTE

*Preach not to others what they should eat, but eat as becomes you and
be silent.*

—*Epictetus*

In the 1965 book *Let's Get Well*, nutritionist Adelle Davis wrote, "We are
indeed much more than what we eat, but what we eat can nevertheless help us
to be much more than what we are."[1] Even after having most of my previous
beliefs about how I thought I should eat to be more sophisticated, attractive,
healthy, and morally responsible challenged, I admit to finding this sentiment attractive. It must speak to the middle-class striver in me. However,
Davis's own legacy stands as a caution against putting too much faith in the
transformational power of food. For a brief period in the 1970s, Davis was the
most recognizable nutritional expert in the country. In a series of best-selling
books and media appearances, she made the case that exercise, a whole-food
diet, and vitamin supplements could cure a wide range of diseases and social
ills, including crime, divorce, and suicide. The scientific citations her books
were full of turned out to be mostly bunk. A review that looked at the 170
sources cited in one of her book chapters found that only 30 accurately supported her assertions.[2]

For the most part, Davis's advice was probably harmless, even if incorrect.
For example, she claimed that she had never known anyone to get cancer if
they drank a quart of milk a day. Then she herself was diagnosed at age sixty-
nine with a multiple myeloma that killed her the following year. Although her
milk habit clearly didn't save her from cancer, it probably didn't play any role
in her death. However, several children whose parents followed the advice in

her books to treat issues such as colic with potassium or vitamin A supplements developed serious illnesses, and one died. Several parents brought lawsuits against Davis's estate and publishers that were settled out of court.[3]

Usually investing in the ideology that food can make us "more than we are" doesn't result in sick or dying children, but it can cause subtler forms of harm. People spend billions of dollars every year on gourmet foods that taste no better, diet foods that don't make them thinner, natural foods that may actually be worse for the environment, and ethnic foods whose real distinguishing feature is merely being relatively difficult to obtain. However misguided, these attempts to eat better may provide real pleasures for many people and answer real needs created by the long decline in the middle-class share of national income. But they have also absorbed time, energy, and money that might have otherwise been invested in more effective or meaningful ways of improving people's lives and communities.

As people including environmental studies professor Michael Maniates and sociology professor Andrew Szasz have argued, individualistic and consumeristic responses to problems such as global warming undermine the collective and structural solutions those problems demand.[4] The possibility of distinguishing yourself by eating the right things probably also predisposes people to believe in particular ways of framing problems that lend themselves to individualistic and consumeristic solutions. For example, people who believe that choosing not to eat certain foods will make them thin may be more likely to believe that Americans in general have gotten dangerously fat by overconsuming those foods so that their own abstention makes them virtuous. People who appreciate the aesthetics of a farmers' market may be more likely to believe the story that locally grown foods typically require less energy to transport and are grown using less toxic chemicals because it justifies their preexisting affinity for shopping there. If those narratives are false, things such as the war on obesity and locavorism may not only waste individuals' time and money but also inspire unwise investments of social resources in problems that might be exaggerated or entirely invented.

Furthermore, the popularization and justification of elite tastes and morals undoubtedly serves to reinforce class hierarchies and reduce social mobility. Like all forms of cultural capital, culinary capital isn't just a way for classes to express value-neutral taste preferences. The food revolution has helped stigmatize the foods and bodies associated with the poor and has convinced middle- and upper-class people that their dietary choices prove that they are smarter and more self-controlled and thus deserve whatever social rewards they get from eating the way they do. Eating and encouraging other people to eat "higher quality" foods may not seem like bigotry, especially when it is justified by concerns about health and the environment, but it works to rein-force pernicious social divides.

For all the discussion about fighting inequality spurred by the Occupy movement and Bernie Sanders's unexpectedly successful challenge to Hillary Clinton's 2016 presidential campaign, there has been little acknowledgment of the ways class hierarchies would persist even if economic inequality were reduced or eliminated. Simply putting more money in the pockets of working-class Americans will not necessarily make them want to eat like the professional middle class that dominates the discourse about food today. That would likely be true even if the food preferences of the professional middle class were clearly superior. Tastes, habits, and norms are ultimately matters of acculturation, and the incentives and opportunities to adopt new cultural practices are not equally available to everyone. Becoming someone who eats according to the ideals of the food revolution is as much a matter of being seen as authentic if you do so as it is of having access to the proper foods.

What's a typical middle-class striver who genuinely likes kale and craft beer but doesn't want to contribute to the calcification of class divides to do? Well, given that social class divisions are to some extent a matter of cultural differences, there may be no way to completely avoid contributing to inequal-ity. No one can avoid being acculturated, which is as true for middle-class strivers as it is for people who still turn their nose up at stinky cheese and foreign food. The more people cultivate the tastes associated with the food revolution—buying groceries, patronizing restaurants, consuming and pro-ducing media that portray a particular set of foods as superior and others as inferior—the more taken for granted they will become and the more the dominant culture will see other tastes as abnormal and distasteful. However, there are a few things that might at least soften the divisive power of the food revolution.

First, people could be more critical about the narratives they embrace when it comes to what kinds of foods are better or worse. When you see foods portrayed as gourmet, healthy, natural, or authentic, try taking a step back and thinking about why those foods have earned that status. Who says they taste better? Is there any evidence of improved health outcomes? What are the ecological costs? What makes them more "authentic"? Few people have the time or ability to exhaustively research every consumption choice they make, but they can at least question the claims they encounter, try to check their sources, and be particularly cautious about too easily embracing stories that merely justify their preexisting preferences.

Second, people who wish to reform the food system could try to avoid tac-tics that presume to dictate to people how they should eat without their input or rely on the "if you build it, they will come" approach. Instead of assuming that poor people want and need community gardens and cooking classes, reform efforts should start by seeking the input of their target populations about what challenges, if any, they face in being able to eat the way they want to eat, not the way reformers think they should eat. Reformers should also be mindful of how status shapes people's ability and desire to adopt new practices. The rewards of consuming high-status foods are not equally available to all, and those who eat food associated with a higher class than their own may receive social censure outside the professional middle class.

Last, people could aim to cultivate a greater awareness and tolerance about how class differences shape taste. To the extent that elite food reformers address the issue of how the poor eat, they tend to focus on the issues of access and education. The first goal is typically to ensure that the poor can eat like the reformers think everyone should. When it turns out that many poor people don't want to eat that way, the reformers aim to find ways to teach them why they should eat. Sometimes there is a nod to providing access to "culturally appropriate" food, particularly when the target population is nonwhite. Nonetheless, the way the reformers talk about the kind of food they want to provide almost always reflects their own beliefs about what is delicious, healthy, appropriate, and normal to eat. These norms are often implicitly and sometimes explicitly derogatory about foods that must be seen as "culturally appropriate" in other classes. If they were not, reformers would not have to spend so much effort trying to get people to stop eating them. This approach essentially treats the tastes of lower classes as broken or defective. In an age when so many people claim with passionate certainty to know how we should eat, taking a more relativist approach and refraining from passing judgment on other people's taste might be what's truly revolutionary.

Acknowledgments

This book began as a dissertation that I started writing during the summer of 2008, when Barack Obama was campaigning for president the first time. As I submit the final draft, his second term in office is coming to a close. In those eight years, many people helped bring this project to fruition.

Paul Allen Anderson first suggested that instead of leaving graduate school with the vague hope of becoming a sommelier, perhaps I should write a dissertation about wine. His support for the project that grew out of that inquiry far exceeded the obligations of a graduate school advisor or dissertation chair. Warren Belasco volunteered to serve as an outside reader and also far overdelivered, providing generous, thoughtful, extensive feedback on innumerable conference presentations and chapter drafts and many kind words of encouragement. My two other committee members, Susan Douglas and Penny Von Eschen, helped shape my thinking from the very beginning of my graduate coursework and offered invaluable suggestions on the dissertation. I hope this book does some credit to their collective brilliance.

Leslie Mitchner at Rutgers University Press has been a great advocate for the project since long before I completed the dissertation, and she patiently shepherded me through the long process of trying to make it accessible and relevant to a broader audience. One anonymous reviewer and Charlotte Biltekoff were kind enough to see the potential in a very imperfect manuscript and offer a wealth of suggestions to help me refine it and clarify its place in the growing body of food scholarship. Kate Babbitt's careful editing and thoughtful suggestions have helped me try to say what I mean more clearly and cite my sources more accurately. I'm also grateful to the many people who offered feedback on drafts, especially Paul Barron, Zoe Stahl, Jesse Carr, Sarah Conrad Gothie, David Denny, Natalie Knazik, Emma Garrett, Kelly

Sisson-Lessens, Susanna Linsley, Sebastian Ferrari, Manan Desai, Kayte Steinbock, Michael d'Emic, Dana Jackson, Mickey Rinaldo, Magdalena Zaborowska, Erik Morales, Wendy Michaels, Annah MacKenzie, Matt Stiffler, Victor Mendoza, Candace Moore, and the members of the Fall 2009 Feeling Theory seminar.

I received financial support from the Program in American Culture, the Center for the Education of Women and Mary Malcomson Raphael Fellowship Fund, the Center for Ethics in Public Life, the Sweetland Writing Center, and the Rackham Graduate School, all at the University of Michigan. I am deeply indebted to the support staff and administrators in the Program in American Culture, especially Marlene Moore, the Department of Communications, and the Sweetland Writing Institute for help in navigating the institutional bureaucracy and for innumerable acts of kindness, large and small. I am also grateful to Phil Deloria, Evans Young, and everyone in the Office of Undergraduate Education who has given me the opportunity to create and teach classes I love. Many brilliant Michigan students have helped me broaden and refine my thinking about how and why people eat the way they do.

The Association for the Study of Food and Society has provided invaluable opportunities to share my work and get feedback from some of the most accomplished scholars working on food. I am especially grateful for the mentorship, support, and friendship of Warren Belasco (again), Ken Albala, Alice Julier, Krishnendu Ray, Doug Constance, Bob Valgenti, Christy Spackman, Jake Lahne, Shawn Trivette, and Bryan Moe.

This book would not exist without the love and support of my friends and family, especially my husband, Brian, who has tried valiantly to break me of some of the dreadful writing habits I picked up in graduate school. The author photo was generously provided by Linda Wan with scanning assistance from Ross Orr. Finally, I owe every success to my parents, who taught me to read, write, cook, and love.

Notes

INTRODUCTION

1. Wine Institute/Gomberg, Fredrikson & Associates, "Wine Consumption in the U.S.," *Wine Institute*, https://www.wineinstitute.org/resources/statistics/article86.

2. Sometimes referred to as "critter wines." Rob Walker, "Animal Pragmatism," *New York Times*, April 23, 2006, accessed September 2, 2015, http://www.nytimes.com/2006/04/23/magazine/23wwln_consumed.html.

3. Steven Shapin, "Hedonistic Fruit Bombs," *London Review of Books* 27, no. 3 (February 2005), accessed September 2, 2015, http://www.lrb.co.uk/v27/n03/steven-shapin/hedonistic-fruit-bombs.

4. Barb Stuckey, *Taste What You're Missing: The Passionate Eater's Guide to Why Good Food Tastes Good* (New York: Free Press, 2012), 160–161.

5. Linda M. Bartoshuk, "Sweetness: History, Preference, and Genetic Variability," *Food Technology* 45, no. 11 (1991): 108–113.

6. Port, sherry, vermouth, and champagne were also available in the form of expensive imports, but they were not nearly as popular or as widely available as the cheap grain alcohol and brandy blends.

7. Kevin Zraly, *American Wine Guide* (New York: Sterling, 2006), 38.

8. Table wines are made by fermenting the juice of grapes or other fruits without added sugar. Wine Institute/Gomberg, Frederickson & Associates, "Wine Consumption in the U.S."

9. "Dago" is an ethnic slur that originally derived from the name Diego and was used as a disparaging term for any foreigner. By the nineteenth and twentieth centuries, it was primarily used to refer to Italian immigrants. *OED Online*, s.v. "Dago, n." accessed September 2, 2015.

10. George Gallup, *The Gallup Poll, 1935–1971* (Wilmington, DE: Scholarly Publishers, 1972), 636, quoted in Harvey Levenstein, *Paradox of Plenty: A Social History of Eating in Modern America* (Berkeley: University of California Press, 1993), 119.

11. Wine Institute/Gomberg, Frederickson & Associates, "Wine Consumption in the U.S."

12. Donna R. Gabaccia, *We Are What We Eat: Ethnic Food and the Making of Americans* (Cambridge, MA: Harvard University Press, 2000), 124.

13. "Alcohol and Drinking," Gallup, accessed September 2, 2105, http://www.gallup.com/poll/1582/alcohol-drinking.aspx.

14. Microbrews are defined as beer produced in the United States by firms with annual production levels between 5,000 and 10,000 barrels. Martin H. Stack, "A Concise History of America's Brewing Industry," EH.net, ed. Robert Whaples, last modified July 4, 2003, https://eh.net/encyclopedia/ a-concise-history-of-americas-brewing-industry.

15. Mrs. G. Edgar Hackey, *Dining for Moderns with Menus and Recipes: The Why and When of Wining* (New York: New York Women's Exchange, 1940), 6, quoted in Megan J. Elias, *Food in the United States, 1890–1945* (Santa Barbara, CA: Greenwood Press, 2009), 38. Women's exchanges were commercial operations run by and for wealthy women who needed to earn money but wanted to avoid the stigma of working for wages or entering the male-dominated world of business. Kathleen Waters Sander, *The Business of Charity: The Woman's Exchange Movement, 1832–1900* (Urbana: University of Illinois Press, 1998).

16. Michelle M. Lelwica, *The Religion of Thinness: Satisfying the Spiritual Hungers behind Women's Obsession with Food and Weight* (Carlsbad, CA: Gürze Books, 2010), 1.

17. There is no one definition of real food. This term and others such as better, bad, good, gourmet, natural, healthy, and authentic are all contested social constructs and should be read with implicit scare quotes.

18. Michael Pollan, *Food Rules: An Eater's Manifesto* (New York: Penguin, 2009), 2.

19. From 1960 to 2000, the weight of the average American adult increased by about twenty pounds and height increased by about an inch. There has been no increase since 2000. The increase pushed many more people over the BMI thresholds for overweight and obesity, but it may not be justified as a cause of concern for public health. Harriet Brown, *Body of Truth: How Science, History, and Culture Drive Our Obsession with Weight—and What We Can Do About It* (Boston: Da Capo Press, 2015), 12. See chapter 3 below for a longer discussion of the evidence about the relationships between fatness, weight-loss dieting, and health.

20. Marty Strange, *Family Farming: A New Economic Vision* (Lincoln: University of Nebraska Press, 1988); and Mary Weaks-Baxter, *Reclaiming the American Farmer: The Reinvention of a Regional Mythology in Twentieth-Century Southern Writing* (Baton Rouge: Louisiana State University Press, 2006).

21. "Estimates of Foodborne Illness in the United States," *Centers for Disease Control and Prevention*, last updated January 8, 2014, http://www.cdc.gov/foodborneburden.

22. Craig Baker-Austin, Joaquin A. Trinanes, Nick G. H. Taylor, Rachel Hartnell, Anja Siitonen, and Jaime Martinez-Urtaza, "Emerging *Vibrio* Risk at High Latitudes in Response to Ocean Warming," *Nature Climate Change* 3 (July 2012): 73–77.

23. From 1990 to 2006, the average life expectancy increased by 3.6 years for American men and 1.9 years for women. The age-adjusted death rate in 2005 was 45 percent lower than in 1950, largely due to declines in mortality from heart disease and stroke. In age-adjusted terms, cancer rates also fell from 475.4 cases per 100,000 people in 1990 to 442.7 in 2005. Michael Gard, *The End of the Obesity Epidemic* (New York: Routledge, 2011), 24 and 72–74.

24. Evadnie Rampersaud, B. D. Mitchell, T. I. Pollin, M. Fu, H. Shen, J. R. O'Connell, J. L. Ducharme, S. Hines, P. Sack, R. Naglieri, A. R. Shuldiner, and S. Snitker, "Physical Activity and the Association of Common FTO Gene Variants with Body Mass Index

and Obesity," *Archives of Internal Medicine* 168, no. 16 (2008): 1791–1797; and Wen-Chi Hsueh, B. D. Mitchell, J. L. Schneider, P. L. St. Jean, T. I. Pollin, M. G. Ehm, M. J. Wagner, D. K. Burns, H. Sakul, C. J. Bell, and A. R. Shuldiner, "Genome-Wide Scan of Obesity in the Old Order Amish," *Journal of Clinical Endocrinology and Metabolism* 86, no. 3 (2001): 1199–1205.

25. Amy Bentley, "Islands of Serenity: The Icon of the Ordered Meal in World War II," in *Food and Culture in the United States: A Reader*, ed. Carol Counihan (New York: Routledge, 2002), 171–192.

26. Warren Belasco, *Appetite for Change: How the Counterculture Took on the Food Industry* (Ithaca, NY: Cornell University Press, 1989), 3.

27. See the discussion of the Gibson girl and flapper in chapter 2.

28. To take just one illustrative year, cookbook sales rose more than 4 percent from 2009 to 2010, while industrywide book sales declined by 4.5 percent, according to Nielsen BookScan. Alice L. McLean, "Cookbooks and the Publishing Industry," in *The SAGE Encyclopedia of Food Issues*, ed. Ken Albala (Los Angeles: Sage Publishers, 2015), 296–299.

29. This use of the term ideology in the humanities is usually attributed to Karl Marx and Friedrich Engels, who argued in *The German Ideology* that the ruling ideas of any given epoch are controlled by the same class that controls the primary means of production. Louis Althusser defined ideology as "a system (with its own logic and rigor) of representations (images, myths, ideas or concepts, depending on the case) endowed with a historical existence and role within a given society," and "the lived relation between men and their world . . . that only appears as 'conscious' on the condition that it is unconscious." Karl Marx and Friedrich Engels, "*The German Ideology*, Part I (selections)," in *Karl Marx: Selected Writings*, ed. Lawrence H. Simon (1845; repr., Indianapolis: Hackett, 1994), 129. See also Louis Althusser, *For Marx*, trans. Ben Brewster (1965; repr., New York: Pantheon Press, 1969), 227–231.

30. Opinion surveys can sometimes offer clues, but it's difficult to ask people directly about beliefs they may not even be aware they have.

31. My focus on both content analysis and audience responses is influenced by Stuart Hall's argument that "the moments of 'encoding' and 'decoding,' though only 'relatively autonomous' in relation to the communicative process as a whole, are *determinate* moments." "Encoding/Decoding," in *Culture, Media, Language: Working Papers in Cultural Studies, 1972–1979*, ed. Stuart Hall, Dorothy Hobson, Andrew Lowe, and Paul Willis (London: Unwin Hyman, Ltd., 1980), 128–138.

32. Steven Poole, *You Aren't What You Eat: Fed Up with Gastroculture* (Toronto, ON: McClelland and Stewart, 2012), 1.

33. The first season premiere of *MasterChef Australia* was the most watched show in its time slot, and the finale was the most watched television broadcast of the year in 2009. The series is the fourth most watched program since OzTAM began tracking ratings in 2001, behind only the 2005 Australian Open final and the 2003 World Rugby Cup final. David Knox, "3.74M Viewers Power *Masterchef Final*," *TV Tonight*, July 20, 2009, http://www.tvtonight.com.au/2009/07/3-74m-viewers-powers-masterchef-finale.html.

34. See the description of Toronto's food scene in Josée Johnston and Shyon Baumann, *Foodies: Democracy and Distinction in the Gourmet Foodscape* (New York: Routledge, 2010), vii.

35. Afua Hirsch, "A Foodie Revolution Cooking in West Africa," *The Guardian*, April 18, 2013, http://www.theguardian.com/world/2013/apr/18/foodie-revolution-cooking-in-west-africa.

36. "Growing Income Inequality in OECD Countries: What Drives It and How Can Policy Tackle It?" OECD Forum on Tackling Inequality, Paris, France, May 2, 2011, http://www.oecd.org/els/soc/47723414.pdf.

37. William Thompson and Joseph Hickey, *Society in Focus* (Boston: Pearson, 2005), 2.

38. Barbara Ehrenreich, *Fear of Falling: The Inner Life of the Middle Class* (New York: Pantheon Books, 1989), 6.

39. Barry Glassner, *The Gospel of Food: Everything You Think You Know about Food Is Wrong* (New York: HarperCollins, 2007), 225.

40. Charlotte Biltekoff, *Eating Right in America: The Cultural Politics of Food and Health* (Durham, NC: Duke University Press, 2013), 9.

41. Helen Zoe Veit, *Modern Food, Moral Food: Self-Control, Science, and the Rise of Modern American Eating in the Early Twentieth Century* (Chapel Hill: University of North Carolina Press, 2013), 185.

42. Andrew P. Haley, *Turning the Tables: Restaurants and the Rise of the American Middle Class, 1880–1920* (Chapel Hill: University of North Carolina Press, 2011), 8–9.

43. Johnston and Baumann, *Foodies*, x.

44. Peter Naccarato and Kathleen LeBesco, *Culinary Capital* (New York: Berg, 2012), 18.

45. Aaron Bobrow-Strain, *White Bread: A Social History of the Store-Bought Loaf* (Boston: Beacon Press, 2012), 15–16.

CHAPTER 1 — INCOMPATIBLE STANDARDS

1. Simon Romero and Sara Shahriari, "Quinoa's Global Success Creates Quandary at Home," *New York Times*, March 19, 2011, http://www.nytimes.com/2011/03/20/world/americas/20bolivia.html.

2. A 2007 laboratory analysis of the nine-course tasting menu at Per Se and McDonald's Big Mac concluded that the nine courses alone were equivalent to two and a half burgers or 1,230.8 Kcal and the amuse bouche, dinner roll, wine, and complementary chocolates brought the total up to 2,416.2 Kcal, or four and a half burgers. Charles Stuart Platkin, "Per Se, per Calorie," *New York Magazine*, June 19, 2007, accessed July 3, 2016, http://nymag.com/restaurants/features/31268.

3. Eddie Gehman Kohan, *Obama Foodorama*, November 4, 2008, accessed September 28, 2013, http://obamafoodorama.blogspot.com.

4. Eddie Gehman Kohan, e-mail message to the author, March 11, 2009.

5. Jocelyn Noveck, "Capital Culture: World Hangs on Obama's Every Bite," *DailyComet.com*, June 8, 2009, accessed July 3, 2016, http://www.dailycomet.com/article/20090613/ARTICLES/906089906?Title=CAPITAL-CULTURE-World-hangs-on-Obama-s-every-bite.

6. "Introduction: *White House Cook Book*," *Feeding America: The Historic American Cookbook Project*, May 21, 2004, http://digital.lib.msu.edu/projects/cookbooks/html/books/book_40.cfm.

7. Maureen Dowd, "'I'm President,' So No More Broccoli!" *New York Times*, March 23, 1990, http://www.nytimes.com/1990/03/23/us/i-m-president-so-no-more-broccoli.html.

8. Katharine Q. Seelye, "Live Blog: The Inauguration of Barak Obama," *The Caucus: The Politics and Government Blog of the New York Times*, January, 20, 2009, http://thecaucus.blogs.nytimes.com/2009/01/20/live-blog-the-inauguration-of-barack-obama.

9. Kim Severson, "How Caramel Developed a Taste for Salt," *New York Times*, December 31, 2008, http://www.nytimes.com/2008/12/31/dining/31cara.html.

10. Jeff Zeleny, "Obama's Down on the Farm," *The Caucus: The Politics and Government Blog of the New York Times*, July 27, 2007, http://thecaucus.blogs.nytimes.com/2007/07/27/obamas-down-on-the-farm.

11. Arugula is listed among the exotics that became popular between 1980 and 1985 along with things such as kiwi, blood oranges, lemongrass, and Tahitian vanilla. Silvia Lovegren, "Historical Overview: From the 1960s to the Present," in *The Oxford Encyclopedia of Food and Drink in America*, ed. Andrew F. Smith, accessed September 2, 2015.

12. John K. Wilson, "George Will, Man of the People, Misquotes Obama, Pushing the Elitist Tag," *Huffington Post*, April 22, 2008, http://www.huffingtonpost.com/john-k-wilson/george-will-man-of-the-pe_b_96506.html.

13. John Fund, "Obama's Flaws Multiply," *Wall Street Journal*, April 15, 2008, http://www.wsj.com/articles/SB120821921853714665.

14. See-Dubya, "Flashback: More Obamessiah Fancy Foodie Follies," *Michelle Malkin*, April 4, 2008, http://michellemalkin.com/2008/04/04/flashback-more-obamessiah-fancy-foodie-follies/; See-Dubya, "Cracker-quiddick Fallout Continues to Haunt SnObama," *Michelle Malkin*, April 15, 2008, http://michellemalkin.com/2008/04/15/cracker-quiddick-fallout-continues-to-haunt-snobama; Michelle Malkin, "Introducing Barack 'Arugula' Obama; Update: CafePress Says, 'No, You Can't,'" *Michelle Malkin*, April 15, 2008, http://michellemalkin.com/2008/04/15/introducing-barack-arugula-obama.

15. Joan Williams, "Obama Eats Arugula," *Huffington Post*, June 9, 2008, http://www.huffingtonpost.com/joan-williams/obama-eats-arugula_b_106166.html.

16. All quotes from online sources are reproduced as published unless otherwise indicated, including departures from standard usage. Comment by Bertha40869, ibid. User comments are no longer available at the original URL but can be seen at https://web.archive.org/web/20080610192051/http://www.huffingtonpost.com/joan-williams/obama-eats-arugula_b_106166.html.

17. Comment by AxelDC in response to ibid.

18. Comments by marla219 and clevelandchick in response to ibid.

19. John McCormick, "Obama Talks Arugula—Again—in Iowa," *The Swamp Blog/ Baltimore Sun*, October 5, 2007, accessed February 28, 2011, http://weblogs.baltimoresun.com/news/politics/blog/2007/10/obama_talks_arugula_again_in_i.html.

20. Carrie Budoff Brown, "Pennsylvania, Meet Barack Obama," *Politico*, March 29, 2008, http://www.politico.com/story/2008/03/pennsylvania-meet-barack-obama-009260.

21. Jon Meacham, "The Editor's Desk," *Newsweek*, May 5, 2008, 4.

22. Nick Kindelsperger, "Obama Makes Surprise Visit at Topolobampo," *Grub Street Chicago*, November 1, 2010, http://www.grubstreet.com/2010/11/obama_makes_surprise_visit_at.html.

23. Bruce Horovitz, "Obama Family Favorites Likely to Get Brand Boost," *USA Today*, November 6, 2008 B.4.

24. Mark Dolliver, "A Few of Barack Obama's Favorite Things," *AdWeek*, May 29, 2008, http://www.adweek.com/adfreak/few-barack-obamas-favorite-things-15809.

25. Paul Bedard, "A Tour of the White House Menu: What Makes Obama a Gourmet President," *U.S. News and World Report*, September 3, 2010, http://www.usnews.com/news/blogs/washington-whispers/2010/09/03/a-tour-of-the-white-house-menu-what-makes-obama-a-gourmet-president.

26. Eric Hananoki, "Dijon Derangement Syndrome: Conservative Media Attack Obama for Burger Order," *Media Matters for America*, May 7, 2009, http://mediamatters.org/research/2009/05/07/dijon-derangement-syndrome-conservative-media-a/149946.

27. Bedard, "A Tour of the White House Menu."

28. As Maureen Dowd wrote, "Barack Obama never again wants to be seen as the hoity-toity guy fretting over the price of arugula at Whole Foods"; Maureen Dowd, "Hold the Fries," *New York Times*, June 17, 2009, http://www.nytimes.com/2009/06/17/opinion/17dowd.html.

29. A website known primarily for catering to "adventurous" eaters that Jim Leff started as a message board in 1997 was sold to CNET Networks in 2006 and acquired by CBS Interactive in 2008. "The History of CHOW," *Chow*, accessed September 2, 2015, http://www.chow.com/about.

30. Lemoncaper, "How Would You Define Gourmet and . . ." *Chowhound*, January 1, 2009, http://chowhound.chow.com/topics/584373, and comment by Kajikit in response.

31. For an example of another time when the ideal of sophistication was more prevalent and populism in retreat, see chapter 2.

32. John Crawley, "The Impact of Obesity on Wages," *Journal of Human Resources* 39, no. 2 (2004): 451–474.

33. Roberta R. Friedman and Rebecca M. Puhl, *Weight Bias: A Social Justice Issue*, Yale Rudd Center for Food Policy and Obesity Policy Brief (2012), accessed September 2, 2015, http://www.uconnruddcenter.org/files/Pdfs/Rudd_Policy_Brief_Weight_Bias.pdf.

34. Helen Penny and Geoffrey Haddock, "Anti-Fat Prejudice among Children: The 'Mere Proximity' Effect," *Journal of Experimental Social Psychology* 43, no. 4 (2007): 678–683.

35. Washingtonian Staff, "What Do You Think of the May Obama Cover of Washingtonian?" *Capital Comment Blog*, April 23, 2009, http://www.washingtonian.com/blogs/capitalcomment/washingtonian/what-do-you-think-of-the-may-obama-cover-of-washingtonian.php.

36. Bauer Griffin, "Obama Shirtless in Hawaii," *Huffington Post*, January 22, 2009, http://www.huffingtonpost.com/2008/12/22/obama-shirtless-in-hawaii_n_152873.html. Comments by Natalie4Obama and gag, no longer available at the article URL but archived at https://web.archive.org/web/20081225232443/http://www.huffingtonpost.com/2008/12/22/obama-shirtless-in-hawaii_n_152873.html?page=2 and https://web.archive.org/web/20081225232503/http://www.huffingtonpost.com/2008/12/22/obama-shirtless-in-hawaii_n_152873.html?page=6.

37. Carl Campanile, "Arnold Kicks Sand in 'Skinny' Obama's Face," *New York Post*, November 1, 2008, http://nypost.com/2008/11/01/arnold-kicks-sand-in-skinny-obamas-face; Susan Jeffords, *Hard Bodies: Hollywood Masculinity in the Reagan Era* (New Brunswick, NJ: Rutgers University Press, 1993).

38. Amy Chozick, "Too Fit to Be President? Facing an Overweight Electorate, Barack Obama Might Find Low Body Fat a Drawback," *Wall Street Journal*, August 1, 2008, http://www.wsj.com/articles/SB121755336096303089.

39. Beth and Mari, comments in response to ibid., accessed June 18, 2009.

40. Comment by Willie Reigh in response to "Republican Schwarzenegger Says Obama Needs to Bulk Up," *Front Row Washington*, Reuters News Agency, October 31, 2008, http://blogs.reuters.com/talesfromthetrail/2008/10/31/republican-schwarzenegger-says-obama-needs-to-bulk-up.

41. Chozick, "Too Fit to Be President?"

42. Rachel Weiner, "WSJ: Obama May Be Too Thin to Be President," *Huffington Post*, August 9, 2008 accessed February 27, 2011, http://www.huffingtonpost.com/2008/08/01/wsj-obama-may-be-too-thin_n_116266.html and comments in response.

43. Chozick, "Too Fit to Be President?"

44. Kerry Trueman, "How Obama Cheats on Eats at Meet 'n' Greets," *Huffington Post*, January 26, 2009, http://www.huffingtonpost.com/kerry-trueman/how-obama-cheats-on-eats_b_160849.html.

45. Maureen Dowd, "The Hillary Waltz," *New York Times*, April 2, 2008, http://www.nytimes.com/2008/04/02/opinion/02dowd.html.

46. Maureen Dowd, "No Ice Cream, Senator?" *New York Times*, July 13, 2008, http://www.nytimes.com/2008/07/13/opinion/13dowd.html.

47. Alice Waters, "Alice Waters's Open Letter to the Obamas," *Gourmet: Food Politics*, January 15, 2009, http://www.gourmet.com/foodpolitics/2009/01/alice-waters-letter-to-barack-obama.html.

48. Todd Kliman, "Alice Waters Was a Foodie Hero. Now She's the Food Police," *NPR*, January 23, 2009, http://www.npr.org/sections/monkeysee/2009/01/the_limitations_of_the_alice_w.html.

49. Rachel Swarns, "A White House Chef Who Wears Two Hats," *New York Times*, November 3, 2009, http://www.nytimes.com/2009/11/04/dining/04kass.html.

50. Comments by Bodie and Patrick in response to Jeff Nield, "Obama Cites Michael Pollan's Sun-Food Agenda," *Treehugger*, November 3, 2008, http://www.trechugger.com/corporate-responsibility/obama-cites-michael-pollans-sun-food-agenda.html. Obama's citation of Pollan received similar coverage on *Civil Eats*, *Ethicurean*, *Obama Foodorama*, *GreenDaily*, *Capital Press*, and *Chow*.

51. Bonnie McCarvel, "MACA Letter to White House about Organic Garden," *Croplife*, March 31, 2009, http://www.croplife.com/crop-inputs/maca-letter-to-white-house-about-organic-garden.

52. Waters, "Alice Waters's Open Letter to the Obamas."

53. Ta-Nehisi Coates, "More on Honest Tea and Obama," *The Atlantic*, August 5, 2008, http://www.theatlantic.com/entertainment/archive/2008/08/more-on-honest-tea-and-obama/5577.

54. Barack Obama, *The Audacity of Hope: Thoughts on Reclaiming the American Dream* (New York: Random House, 2006), 249.

55. Todd Kliman, "Barack Obama: First Eater," *Monkey See: Pop Culture and News Analysis from NPR*, November 6, 2008, http://www.npr.org/sections/monkeysee/2008/11/barack_obama_first_eater.html.

56. Damon Beres, "The Definitive Guide to Obama's Favorite Restaurants," *Esquire*, September 9, 2010, http://www.esquire.com/food-drink/food/a8406/obama-favorite-restaurants.

57. "Obama at Ben's Chili Bowl: Our New President Knows How to Eat," *Chomposaurus: The Meat Blog*, January 13, 2009, https://chomposaurus.wordpress.com/2009/01/13/obama-at-bens-chili-bowl-our-new-president-knows-how-to-eat.

58. Eddie Gehman Kohan, "President-Elect Obama Visits Ben's Chili Bowl in DC . . . A Legendary Joint That's All About Change," *Obama Foodorama*, January 10, 2009, accessed February 22, 2011, http://obamafoodorama.blogspot.com/2009/01/barack-blesses-bens-chili-bowl-in-dca.html.

59. "'Meet the Press' Transcript for January 11, 2009: Cosby, Poussaint, Fenty, Waters, Roundtable," *NBC News*, January 11, 2009, http://www.nbcnews.com/id/28605356/ns/meet_the_press/t/meet-press-transcript-jan.

60. Alicia Villarosa, "The Obama Health Challenge: How Barack and Michelle Can Make the Country Healthier," *The Root*, January 27, 2009, http://www.theroot.com/articles/politics/2009/01/the_obama_health_challenge.html.

61. Eddie Gehman Kohan, "'Politico' Asks Russell Simmons and Ob Fo the Same Question about Obama Folk Foodways. Natch, The Answers Are Wildly Different," *Obama Foodorama*, June 12, 2009, accessed February 12, 2011. http://obamafoodorama.blogspot.com/2009/06/politico-asks-russell-simmons-and-ob-fo.html.

62. Dowd, "Hold the Fries."

63. Severson, "How Caramel Developed a Taste."

64. Paul Campos, *The Obesity Myth: Why America's Obsession with Weight Is Hazardous to Your Health* (New York: Gotham Books, 2004), 68–69.

65. Richard Lawson, "Barack Obama Shames Americans with His Elitist Body," *Gawker*, August 15, 2008, http://gawker.com/5037451/barack-obama-shames-americans-with-his-elitist-body.

66. Seth Koven, *Slumming: Sexual and Social Politics in Victorian London* (Princeton, NJ: Princeton University Press, 2004); Chad Heap, *Slumming: Sexual and Racial Encounters in American Nightlife, 1885–1940* (Chicago: University of Chicago Press, 2009).

67. Warren Belasco, *Appetite for Change: How the Counterculture Took on the Food Industry*, 2nd ed. (1989; repr., New York: Pantheon Press, 2007), 23–24.

68. Ibid., 208.

69. Ibid., 209.

70. Marissa Piesman and Marilee Hartley, *The Yuppie Handbook: The State-of-the-Art Manual for Young Urban Professionals* (New York: Long Shadow Books, 1984). Jacques Chazaud designed the book.

71. According to Brooks, there was a significant shift from 1980s yuppie culture to 1990s Bobos. However, that supposed shift relies on a characterization of the yuppie as considerably more politically conservative and socially straitlaced than the demographic is portrayed in texts like such as *The Yuppie Handbook* and *American Psycho*. If yuppies were not just corporate strivers but instead were also invested in being "hip," tolerant of drug use, and socially liberal, then his argument for the

distinctiveness of "Bobos" becomes much weaker. David Brooks, *Bobos in Paradise: The New Upper Class and How They Got There* (New York: Simon & Schuster, 2000), 11–12.

72. Mark Ames, "Spite the Vote," *New York Press*, June 15, 2004, accessed March 1, 2001, https://newyorkpress.com/spite-the-vote.

73. Critics such as Larry M. Bartels have demonstrated that based on presidential voting behavior, white working-class voters have not migrated to the Republican Party outside the South, where the shift is due entirely to the demise of the South as a bastion of Democratic support in the wake of major shifts on civil rights issues within the Democratic Party in the 1960s. Most white working-class voters claim to place more importance on economic issues than cultural ones and see themselves as closer to the Democratic Party on the so-called cultural wedge issues. Larry M. Bartels, "What's the Matter with *What's the Matter with Kansas?*" *Quarterly Journal of Political Science* 1 (2006): 201–226.

74. Mark Rolfe, "Days of Wine and Poseurs: Stereotypes of Class, Consumption, and Competition in Democratic Discourse," paper presented at the annual meeting of Australasian Political Studies Association, September 24–26, 2007, Monash University, Melbourne, Victoria.

75. John Downing, "Quinn Bids to Banish 'Smoked Salmon Socialist' Image," *Irish Examiner*, March 23, 2002, http://www.irishexaminer.com/archives/2002/0323/ireland/quinn-bids-to-banish-smoked-salmon-socialist-image-25292.html.

76. Jeff Gordinier, "The Return of the Yuppie," *Details*, November 2006, accessed March 1, 2011, http://www.details.com/culture-trends/critical-eye/200611/the-return-of-the-yuppie.

77. Concern about liberal bias in academia and journalism have prompted surveys that generally show that both lean left of the American population as a whole. In a 2007 study of 630,00 full-time professors at institutions ranging from research universities to community colleges, 62.2 percent identified as some kind of liberal and only 19.7 percent as some kind of conservative, compared to the 23.3 percent of the American population that identifies as liberal and the 31.9 percent that identified as conservative. In the 2004 election, 77.6 percent of professors voted for Kerry compared to 48.2 percent of the voting public. Louis Menand, *The Marketplace of Ideas: Reform and Resistance in the American University* (New York: W. W. Norton, 2010), 135–136. According to the Pew Research Journalism Project, 41 percent of journalists identify as left-leaning, compared to 17 percent of the public, and in a poll by the Media Studies Center, 89 percent of Washington journalists said they voted for Clinton in the 1992 election. Eric Alterman, *What Liberal Media? The Truth about Bias and the News* (New York: Basic Books, 2003), 16.

78. Paul Fussell, *Class: A Guide through the American Status System* (New York: Touchstone, 1983), 20.

79. Thorstein Veblen, *The Theory of the Leisure Class* (1899; repr., New York: Random House Modern Library, 1934), 46.

80. Pierre Bourdieu, *Distinction: A Social Critique of the Judgment of Taste*, trans. Richard Nice (1979; repr., Cambridge, MA: Harvard University Press, 1984), 6.

81. Ibid., 35–41.

82. Ibid., 176.

83. Ibid., 7.

84. Stephen Mennell, *All Manners of Food: Eating and Taste in England and France from the Middle Ages to the Present* (Champaign: University of Illinois Press, 1996).

85. Richard Peterson, "Problems in Comparative Research: The Example of Omnivorousness," *Poetics: Journal of Empirical Research on Culture, the Media, and the Arts* 33, nos. 5–6 (2005): 257–282.

86. Peter Naccarato and Kathleen LeBesco, *Culinary Capital* (New York: Berg, 2012), 4.

87. Helen Zoe Veit, *Modern Food Moral Food: Self-Control, Science, and the Rise of Modern American Eating in the Early Twentieth Century* (Chapel Hill: University of North Carolina Press, 2013), 4.

88. Ibid., 5.

89. Ibid., 11.

90. Ibid., 22.

91. Ibid., 23.

92. Bourdieu, *Distinction*, 196.

93. Jennifer Di Noia, "Defining Powerhouse Fruits and Vegetables: A Nutrient Density Approach," *Preventing Chronic Disease* 11 (2014), http://www.cdc.gov/pcd/issues/2014/13_0390.htm.

94. Robert Crawford, "Healthism and the Medicalization of Everyday Life," *International Journal of Health Services: Planning, Administration, and Evaluation* 10, no. 3 (1980): 365–388.

95. K. J. A. Colasanti, David S. Connor, and Susan B. Smalley, "Understanding Barriers to Farmers' Market Patronage in Michigan: Perspectives from Marginalized Populations," *Journal of Hunger and Environmental Nutrition* 5, no. 3 (2010): 316–338.

96. Reay Tannahill, *Food in History* (New York: Stein and Day, 1973), 393, quoted in Nacarrato and LeBesco, *Culinary Capital*, 9.

CHAPTER 2 — ASPIRATIONAL EATING

1. Richard Ohmann argues that advertising-driven magazines, which were much less expensive and aimed at a slightly less educated and affluent audience than older subscription-driven magazines, were the first example of national mass culture in the United States and that they succeeded because they responded to the changing needs of industrial capitalism, particularly the need to cultivate consumption. *Selling Culture: Magazines, Markets, and Class at the Turn of the Century* (New York: Verso, 1996).

2. Amy Bentley refers to this as the A + 2B meal, one dominated by a central high-status protein, ideally beef, flanked by two less-prized starch or vegetable sides. *Eating for Victory: Food Rationing and the Politics of Domesticity* (Champaign: University of Illinois Press, 1998).

3. Charles W. Calhoun, "The Political Culture: Public Life and the Conduct of Politics," in *The Gilded Age: Perspectives on the Origins of Modern America*, ed. Charles W. Calhoun (Lanham, MD: Rowman & Littlefield, 2007), 240.

4. Charles W. Calhoun, "Introduction," in *The Gilded Age: Perspectives on the Origins of Modern America*, ed. Charles W. Calhoun (Lanham, MD: Rowman & Littlefield, 2007), 2.

5. Eric Arnesen, "American Workers and the Labor Movement in the Late Nineteenth Century," in *The Gilded Age: Perspectives on the Origins of Modern America*, ed. Charles W. Calhoun (Lanham, MD: Rowman & Littlefield, 2007), 56.

6. Lawrence Levine, *Highbrow/Lowbrow: The Emergence of Cultural Hierarchy in America* (Cambridge, MA: Harvard University Press, 1988).

7. *Service à la française* referred to the style of service where servants brought in the dishes and then left while the guests served themselves. Ed Crews, "Thomas Jefferson: Culinary Revolutionary," *Colonial Williamsburg Journal* (Winter 2013), https://www.history.org/Foundation/journal/summer13/jefferson.cfm.

8. Charles Ranhofer, *The Epicurean: A Complete Treatise of Analytical and Practical Studies on the Culinary Art, including Table and Wine Service, How to Prepare and Cook Dishes . . . etc., and a Selection of Interesting Bills of Fare from Delmonico's from 1862 to 1894* (New York: Charles Ranhofer, 1894), 8.

9. Ellen Koteff, "Thomas Jefferson: America's First Gourmet," *Nation's Restaurant News*, February 1996.

10. Harvey Levenstein, *Revolution at the Table: The Transformation of the American Diet* (New York: Oxford University Press, 1988), 11.

11. Ibid.

12. Catherine E. Beecher and Harriet Beecher Stowe, *The American Woman's Home; or, Principles of Domestic Science* (New York: J. B. Ford & Co., 1869), 179.

13. Kristin Hoganson, *Consumer's Imperium: The Global Production of American Domesticity, 1865–1920* (Chapel Hill: University of North Carolina Press, 2007), 106.

14. Levenstein, *Revolution at the Table*, 8.

15. Ibid., 15.

16. Hoganson, *Consumer's Imperium*, 106.

17. Maria Parola, *Miss Parola's Kitchen Companion: A Guide for All Who Would Be Good Housekeepers* (Boston: Estes and Lauriat, 1887), 910–911.

18. Levenstein, *Revolution at the Table*, 20.

19. Ranhofer, *The Epicurean*, 8.

20. Levenstein, *Revolution at the Table*, 20.

21. Mrs. John M. E. W. Sherwood, *Manners and Social Usages* (1887; repr., New York: Harper & Brothers, 1907), 121.

22. Arthur Van Vlissingen Jr., "Fred Harvey: Applying Factory Methods to Serving Meals," *Factory and Industrial Management* 78 (November 1929): 1083–1085.

23. Harvey Levenstein, *Paradox of Plenty: A Social History of Eating in Modern America* (Berkeley: University of California Press, 1993), 45.

24. Hillel Schwartz, *Never Satisfied: A Cultural History of Diets, Fantasies, and Fat* (New York: Anchor Books, 1986), 21.

25. Ibid.

26. Valerie Steele, *Fashion and Eroticism: Ideals of Feminine Beauty from the Victorian Era to the Jazz Age* (New York: Oxford University Press, 1985), 108.

27. A Professional Beauty, *Beauty and How to Keep It* (London: Brentano's, 1889), 47, quoted in ibid.

28. Armond Fields, *A Biography of America's Beauty* (Jefferson, NC: McFarland & Co., 1999), 85.

29. Ibid., 98.

30. Lois Banner, *American Beauty: A Social History* (New York: Alfred A. Knopf, 1983), 151.

31. Lillian Russell, "Lillian Russell's Beauty Secrets: Do You Eat Properly?" *Chicago Daily Tribune*, August 7, 1911, 4.

32. Banner, *American Beauty*, 151.

33. Joan Jacobs Brumberg, *Fasting Girls: A History of Anorexia Nervosa* (1988; repr., New York: Vintage Books, 2000), 182.

34. Rebecca Spang, *The Invention of the Restaurant: Paris and Modern Gastronomic Culture* (Cambridge, MA: Harvard University Press, 2000).

35. Brumberg, *Fasting Girls*, 184.

36. Angelika Köhler, "Charged with Ambiguity: The Image of the New Woman in American Cartoons," in *New Woman Hybridities: Femininity, Feminism, and Consumer Culture, 1880–1930*, ed. Ann Heilmann and Margaret Bettham (New York: Routledge, 2004), 158–178.

37. Angela J. Latham, *Posing a Threat: Flappers, Chorus Girls, and Other Brazen Performers of the American 1920s* (Hanover, NH: Wesleyan University Press, 2000), 10.

38. Aldous Huxley, *Antic Hay* (1923; Normal, IL: Dalkey Archive Press, 1997), 82, quoted in Valerie Steele, *The Corset: A Cultural History* (New Haven, CT: Yale University Press, 2003), 154.

39. Laura Fraser, *Losing It: False Hopes and Fat Profits in the Diet Industry* (New York: Plume, 1998), 16.

40. Ibid., 17.

41. Ibid.

42. Schwartz, *Never Satisfied*, 204.

43. Ibid.,113. Slow chewing was often called "Fletcherism," after one of its primary advocates. There were earlier "mastication" diets aimed at alleviating indigestion, but Fletcher's "industrious chewing" was focused on the time involved in chewing, not the mechanical process of breaking down food. According to Schwartz, it caught on because it "made the eating of less food into the formidable pretense of eating more" (ibid., 126). Although that may seem strange now, some contemporary dieters echo Fletcher's method. For example, in a 2008 article in her magazine, Oprah says, "in order not to abuse food, I have to stay fully conscious and aware, of every bite, of taking time and chewing slowly." Oprah Winfrey, "How Did I Let This Happen Again?" *Oprah.com*, December 10, 2008, http://www.oprah.com/spirit/Oprahs-Battle-with-Weight-Gain-O-January-2009-Cover.

44. Contrary to popular wisdom, which suggests that the calorie was already used as a measure of heat in the metric system and was adapted to help measure human energy needs and food energy values, James Hargrove argues that "the thermal calorie was not fully defined until the 20th century, by which time the nutritional Calorie was embedded in U.S. popular culture and nutritional policy." James Hargrove, "History of the Calorie in Nutrition," *Journal of Nutrition* 136 (December 2006): 2958.

45. H. I. Phillips, "It Is Never Too Late to Shrink," *American Magazine* 100 (December 1925): 39, quoted in Schwartz, *Never Satisfied*, 183.

46. Fraser, *Losing It*, 19; Schwartz, *Never Satisfied*, 18.

47. Francis G. Benedict, "Rationale of Weight Reduction," *Scientific Monthly* 33 (September 1931): 264, quoted in Schwartz, *Never Satisfied*, 183.

48. J. Eric Oliver, *Fat Politics: The Real Story behind America's Obesity Epidemic* (New York: Oxford University Press, 2005), 70.

49. A. Mazur, "US Trends in Feminine Beauty and Overadaptation," *Journal of Sex Research* 22 (1986): 281–303 and references therein (Laver 1937, Kinsey Pomeroy & Martin 1948, Winch 1952, Morrison 1965, and Jesser 1971).

50. Patricia Vertinsky, "Physique as Destiny: William H. Sheldon, Barbara Honeyman Heath, and the Struggle for Hegemony in the Science of Somatotyping," *Canadian Bulletin of Medical History* 24, no. 2 (2007): 296.

51. Eric Schlosser, *Fast Food Nation: The Dark Side of the American Meal* (New York: Houghton Mifflin, 2001), 153.

52. Gerald A. Danzer, J. Jorge Klor de Alva, Larry S. Krieger, Louis E. Wilson, and Nancy Woloch, *The Americans* (Evanston, IL: McDougal Littell, 2007), 526–528.

53. Lorine Swainston Goodwin, *The Pure Food, Drink, and Drug Crusaders, 1879–1914* (Jefferson, NC: McFarland & Co., 1999).

54. Ibid.

55. Ibid., 15.

56. Ibid.

57. The book was published in serialized form in the socialist newspaper *Appeal to Reason* starting in February 1905, but it never sold much beyond its customary 294,868 subscription copies. The substantial popular press coverage of the novel and its critics all followed the publication of the Doubleday, Page edition. Ronald Gottesman, "Introduction," *The Jungle* (1906; repr., New York: Penguin Books, 2006), xxvi–xxxi.

58. Clayton A. Coppin and Jack High, *The Politics of Purity: Harvey Washington Wiley and the Origins of Federal Food Policy* (Ann Arbor: University of Michigan Press, 1999), 82. Reformers often saw small businesses as especially prone to unclean practices. Donna Gabaccia also claims that many culinary reformers perceived ethnic foods and business as dangerous to the health and well-being of American eaters. She describes the effect of food reform on ethnic entrepreneurs: "Small businessmen experienced regulation as harassment. Reformers explicitly attacked immigrant businessmen as unsanitary when they worked out of their homes. In New York, progressives drove small-scale bakeries and pasta-manufactories out of tenement basements; they legally and precisely specified dimensions for floors, ceilings, and windows of factories, and the quality and location of furnishings, troughs, utensils, and ventilation." Donna Gabaccia, *We Are What We Eat: Ethnic Food and the Making of Americans* (Cambridge, MA: Harvard University Press, 2000), 132.

59. Eric Schlosser, "Preface," *The Jungle* (1906; repr., New York: Penguin Books, 2006), xi.

60. Upton Sinclair, *The Jungle* (1906; repr., New York: Penguin Books, 2006), 378.

61. Richard Hofstadter, *The Age of Reform: From Bryan to F.D.R.* (New York: Vintage, 1955); George E. Mowry, "The California Progressive and His Rationale: A Study in Middle Class Politics," *Mississippi Valley Historical Review* 26 (September 1949): 239–250; Samuel P. Hays, *Conservation and the Gospel of Efficiency: The Progressive Conservation Movement, 1890–1920* (Cambridge, MA: Harvard University Press, 1959); Robert H. Wiebe, *The Search for Order, 1877–1920* (New York: Hill and Wang, 1967); Karen J. Blair, *The Clubwoman as Feminist: True Womanhood Redefined, 1868–1914* (New York: Holmes and Meier, 1980).

62. Goodwin, *The Pure Food, Drink, and Drug Crusaders*, 133.

63. Elizabeth Lindsay Davis, *Lifting as They Climb* (Chicago: National Association of Colored Women, 1933); Willard B. Gatewood, *Aristocrats of Color: The Black Elite, 1880–1920* (Bloomington: Indiana University Press, 1990).

64. Goodwin, *The Pure Food, Drink, and Drug Crusaders*, 266.

65. Hofstadter, *The Age of Reform*; Wiebe, *The Search for Order*; Anne Firor Scott, *The Southern Lady: From Pedestal to Politics, 1830–1930* (Chicago: University of Chicago Press, 1971); Robyn Muncy, *Creating a Female Dominion in American Reform, 1890–1935* (New York: Oxford University Press, 1991); Ian Tyrrell, *Women's World, Women's Empire: The Women's Christian Temperance Union in International Perspective, 1880–1930* (Chapel Hill: University of North Carolina Press, 1991).

66. Hoganson, *Consumer's Imperium*, 105–106.

67. "Middle-Age Cookery," *New York Times*, October 25, 1872, 4, quoted in Andrew P. Haley, *Turning the Tables: Restaurants and the Rise of the American Middle Class, 1880–1920* (Chapel Hill: University of North Carolina Press, 2011), 95.

68. "Home Interests," *New York Tribune*, March 14, 1880, quoted in Hoganson, *Consumer's Imperium*, 107.

69. Hoganson, *Consumer's Imperium*, 136.

70. Martha Esposito Shea and Mike Mathis, *Campbell Soup Company* (Charleston, SC: Arcadia Publishing, 2002), 29.

71. Hoganson, *Consumer's Imperium*, 115.

72. Ibid., 107–108.

73. Karen Halttunen, *Confidence Men and Painted Women: A Study of Middle-Class Culture in America, 1830–1870* (New Haven, CT: Yale University Press, 1982), 101–102.

74. "Chat," *Demorest's Family Magazine* 29 (November 1892): 51, quoted in Hoganson, *Consumer's Imperium*, 141.

75. Hoganson, *Consumer's Imperium*, 141.

76. Ellye Howell Glover, *"Dame Curtsey's" Book of Party Pastimes for the Up-to-Date Hostess* (Chicago: A. C. McClurg & Co., 1912), 211, quoted in Hoganson, *Consumer's Imperium*, 141.

77. Haley, *Turning the Tables*, 97–98.

78. Noah Brooks, "Restaurant Life of San Francisco," *Overland Monthly*, November 1868, 467, quoted in Haley, *Turning the Tables*, 96.

79. Charles Greene, "Restaurants of San Francisco," *Overland Monthly*, December 1892, 566, quoted in Haley, *Turning the Tables*, 101.

80. H. D. Miller, "The Great Sushi Craze of 1905," *An Eccentric Culinary History*, August 10, 2015, http://eccentricculinary.com/the-great-sushi-craze-of-1905-part-2.

81. Gabaccia, *We Are What We Eat*, 100–101.

82. Ibid., 147.

83. Shea and Mathis, *Campbell Soup Company*, 130.

84. June Owen, "How New Yorkers Eat: Rich and Poor Have Monotonous Diets," *New York Times*, March 16, 1961, 40.

85. In the African American neighborhood of Harlem, cooks made use of cheaper cuts of meat and adopted a more casual attitude toward the time of the evening meal, and recent Puerto Rican immigrants reported a preference for rice and bean dishes. Ibid.

86. Ibid.

87. Gabaccia, *We Are What We Eat*, 100.

88. Peter J. Brown estimates on the basis of the Human Relations Area Files that "the desirability of 'plumpness' or being 'filled out' is found in 81 percent of societies for which this variable can be coded." Quoted in Rebecca Popenoe, *Feeding Desire: Fatness, Beauty, and Sexuality Among a Saharan People* (New York: Routledge, 2004), 4.

89. Fraser, *Losing It*, 18.

90. James McWilliams, *A Revolution in Eating: How the Quest for Food Shaped America* (New York: Columbia University Press, 2005), 284.

91. Roger Horowitz, *Putting Meat on the American Table: Taste, Technology, Transformation* (Baltimore, MD: Johns Hopkins University Press, 2006), 12.

92. Schwartz, *Never Satisfied*, 157.

93. Ibid., 170.

94. Ibid., 86.

95. This theory is also echoed by Joseph Roach's argument about performances of waste deflecting the "anxieties produced by having too much of everything—including material goods and human beings"; *Cities of the Dead: Circum-Atlantic Performance* (New York: Columbia University Press, 1996), 123.

96. T. J. Jackson Lears, *Fables of Abundance: A Cultural History of Advertising in America* (New York: Basic Books, 1994), 167–168.

97. Schwartz, *Never Satisfied*, 135.

98. Irving Fisher and Eugene Fisk, *How to Live: Rules for Healthful Living Based on Modern Science* (New York: Funk and Wagnalls, 1915), 33–34.

99. Paula Baker, "The Domestication of Politics: Women and American Political Society, 1780–1920," *American Historical Review* 89, no. 3 (June 1984): 620–647.

100. Karen Blair, *The Clubwoman as Feminist: True Womanhood Redefined, 1868–1914* (New York: Holmes and Meier, 1980), 1.

101. Hoganson, *Consumer's Imperium*, 12.

102. Hofstadter, *The Age of Reform*, 39.

103. According to a table he compiled from reports of the United States Census Bureau, the average death rate per 100,000 for the five years ending with 1904 for influenza was 17.6 in the city and 29.3 in rural communities and for dysentery it was 8.6 in the city and 11.0 in rural communities. Isaac Williams Brewer, *Rural Hygiene* (Philadelphia: J. B. Lippincott, 1909), 13.

104. Peter Barton Hutt and Peter Barton Hutt II, "A History of Government Regulation of Adulteration and Misbranding of Food," *Food Drug and Cosmetic Law Journal* 39 (1984): 2–73; Jack S. Blocker, *American Temperance Movements: Cycles of Reform* (Boston: Twayne Publishers, 1989); Coppin and High, *The Politics of Purity*.

105. See Laura Shapiro, *Something from the Oven: Reinventing Dinner in 1950s America* (New York: Penguin Books, 2004).

106. Claudia Golden and Robert Margo, "The Great Compression: The Wage Structure in the United States at Mid-Century," National Bureau of Economic Research Working Paper no. 3817, August 1991.

107. Alexis de Tocqueville, *Democracy in America*, trans. Arthur Goldhammer (New York: Library of America, 2004), 646, quoted in Thomas Piketty, *Capital in the Twenty-First Century* (Cambridge, MA: Harvard University Press, 2014), 150.

108. Piketty, *Capital in the Twenty-First Century*, 151.

109. Ibid.

110. Levenstein, *The Paradox of Plenty*, 14.

111. Hofstadter, *The Age of Reform*, 137.

112. Davis, *Lifting as They Climb*; and Gatewood, *Aristocrats of Color*.

113. Private wealth declined from nearly five years of national income in 1930 to less than three and a half in 1970. Piketty, *Capital in the Twenty-First Century*, fig. 4.8.

CHAPTER 3 — NO CULINARY ENLIGHTENMENT

1. Barb Stuckey, *Taste What You're Missing: The Passionate Eater's Guide to Why Good Food Tastes Good* (New York: Free Press, 2012), 8–13.

2. "Masters of Disaster," *Top Chef Masters*, season 1, episode 9, aired August 12, 2009, on Bravo.

3. Ibid.

4. Ibid.

5. Sarah Jersild, "Top Chef Masters: No, Spike, It's STILL Not about You," *zap2it*, August 12, 2009, http://zap2it.com/2009/08/top-chef-masters-no-spike-its-still-not-about-you.

6. Krista Simmons, "Top Chef Masters Turns into a Sausage Fest," *LA Times: Daily Dish*, August 13, 2009, http://latimesblogs.latimes.com/dailydish/2009/08/top-chef-masters-and-then-there-were-three.html.

7. David Hume, "On the Standard of Taste," in *English Essays, From Sir Philip Sidney to Macaulay*, ed. Charles W. Eliot (New York: P. F. Collier & Son, 1910), 222.

8. Ibid., 223–228.

9. Ibid., 224.

10. Carl Wilson, *Let's Talk About Love: A Journey to the End of Taste* (New York: Continuum, 2007), 81.

11. Stuckey, *Taste What You're Missing*, 17–19.

12. Ibid., 22–23.

13. Eric Schlosser, *Fast Food Nation: The Dark Side of the All-American Meal* (New York: Harper, 2001), 125.

14. Brian Wansink, *Mindless Eating: Why We Eat More Than We Think* (New York: Bantam, 2006), 120.

15. Tamar Haspel, "Backyard Eggs vs. Store-Bought: They Taste the Same," *Washington Post*, June 2, 2010, http://www.washingtonpost.com/wp-dyn/content/article/2010/06/01/AR2010060100792.html. Haspel's blind tasting of eggs from her backyard hens' eggs and store-bought eggs found no difference. Similarly, a *Serious Eats* panel found that people perceived differences when they could detect visual differences, but not when the differences were obscured with a neutral-tasting food dye. J. Kenji López-Alt, "The Food Lab: Do 'Better' Eggs Really Taste Better?" *Serious Eats*, August 27, 2010, http://www.seriouseats.com/2010/08/what-are-the-best-eggs-cage-free-organic-omega-3s-grocery-store-brand-the-food-lab.html.

16. Debra Zellner, Evan Siemers, Vincenzo Teran, Rebecca Conroy, Mia Lankford, Alexis Agrafiotis, Lisa Ambrose, and Paul Locher, "Neatness Counts. How Plating Affects Liking for the Taste of Food," *Appetite* 57, no. 3 (December 2011): 642–648.

17. Charles Michel, Carlos Velasco, Elia Gatti, and Charles Spence, "A Taste of Kandinsky: Assessing the Influence of the Artistic Visual Presentation of Food on the Dining Experience," *Flavour* 3, no. 7 (June 2014), http://flavourjournal.biomedcentral.com/articles/10.1186/2044-7248-3-7.

18. David Just, Ozge Sigirci and Brian Wansink "Lower Buffet Prices Lead to Less Taste Satisfaction," *Journal of Sensory Studies* 29, no. 5 (October 2014): 362–370.

19. The results are reversed when people are given the information about price or origin after tasting the product instead of before. The researchers predicted the result, suggesting that when information is presented after an affective judgment has been

reached, it results in a "contrast effect" such that consumers evaluate the product more negatively when the information is associated with a favorable experience and more positively when the information is associated with a negative experience. Keith Wilcox, Anne L. Roggeveen, and Dhruv Grewal, "Shall I Tell You Now or Later? Assimilation and Contrast in the Evaluation of Experiential Products," *Journal of Consumer Research* 38, no. 4 (2011): 763–773.

20. Johan Almenberg and Anna Dreber, "When Does the Price Affect the Taste? Results from a Wine Experiment," *Journal of Wine Economics* 6, no. 1 (2011): 111–121; Hilke Plassmann, John O'Doherty, Baba Shiv, and Antonio Rangel, "Marketing Actions Can Modulate Neural Representations of Experienced Pleasantness," *Proceedings of the National Academy of Science* 105, no. 3 (2008): 1050–1054.

21. Ian Sample, "Expensive Wine and Cheap Plonk Taste the Same to Most People," *The Guardian*, April 14, 2011, https://www.theguardian.com/science/2011/apr/14/expensive-wine-cheap-plonk-taste.

22. Robin Goldstein, Johan Almenberg, Anna Dreber, John W. Emerson, Alexis Herschkowitsch, and Jacob Katz, "Do More Expensive Wines Taste Better? Evidence from a Large Sample of Blind Tastings," *Journal of Wine Economics* 3 (Spring 2008): 1–9.

23. Plassmann et al., "Marketing Actions Can Modulate Neural Representations of Experienced Pleasantness," 1050–1054.

24. Descriptors characteristic of red and white wines were determined using a textual analysis program called ALCESTE that counted the co-occurrence of words in particular blocks of text. Brochet ran the analysis on five collections of tasting notes: notes on 100,000 wines from the *Hatchette Guide to the Wines of France*; 3,000 from Jacques Dupont, who published the weekly Gault & Millau letter; 9,000 from Robert Parker; who writes *The Wine Advocate*; 2,000 from Brochet's personal corpus; and 352 from a single blind tasting session of eight wines by forty-four "tasters of international reputation." Although there were few overlaps between the different authors' corpuses—Brochet says no more than ten words were common to more than two authors—each body of tasting notes was strongly indexed by color. Frédéric Brochet and Gil Morrot, "Influence of the Context on the Perception of Wine—Cognitive and Methodological Implications," *Journal International des Sciences de la Vigne et du Vin* 33 (1999): 187–192.

25. The neutrality of the dye, which was made from purified grape anthocyanins, was tested in a separate trial. Fifty people were recruited to taste the white wine with and without the added color served in opaque glasses in booths illuminated by red light to obscure the visual difference. The glasses were presented in random order with three random digits indicated on the glass so neither the experimenter or subject knew which wine contained the red dye. Gil Morrot, Frédéric Brochet, and Denis Dubourdieu, "The Color of Odors," *Brain and Language* 77 (2001): 187–196.

26. Robert T. Hodgson, "An Examination of Judge Reliability at a Major U.S. Wine Competition," *Journal of Wine Economics* 3, no. 2 (2008): 105–113.

27. David Derbyshire, "Wine-Tasting: It's Junk Science," *The Guardian*, June 22, 2013, http://www.theguardian.com/lifeandstyle/2013/jun/23/wine-tasting-junk-science-analysis.

28. Hume, "On the Standard of Taste," 228.

29. There are related debates about the best form of exercise and its aims. In general, proponents of the energy balance paradigm advocate any form of exercise that expends calories. Many low-carb and paleo dieters promote high-intensity exercise and resistance training over consistent cardio because the former may be better at increasing lean muscle mass, which is of greater concern for them than producing a calorie deficit. These debates have been somewhat less contentious, especially in mainstream public discourse, than the ones over diet, perhaps because exercise is often seen as a complement to prudent dietary choices that is insufficient to achieve or maintain thinness on its own.

30. Richard S. Taylor, "Use of Body Mass Index for Monitoring Growth and Obesity," *Paediatrics and Child Health* 15, no. 5 (2010): 258.

31. Joyce B. Harp and Lindsay Hect, "Obesity in the National Football League," *Journal of the American Medical Association* (hereafter *JAMA*) 293, no. 3 (2005): 1058–1062.

32. The National Health and Nutrition Examination Survey includes an interview and examination, which makes it a better source of data on weight than surveys that rely exclusively on self-reports. It has been administered periodically since 1960 to a nationally representative cross-section of the U.S. population. According to the CDC, "The NHANES interview includes demographic, socioeconomic, dietary, and health-related questions. The examination component consists of medical, dental, and physiological measurements, as well as laboratory tests administered by highly trained medical personnel." CDC, "About the National Health and Nutrition Examination Survey," CDC.gov, February 3, 2014, http://www.cdc.gov/nchs/nhanes/about_nhanes.htm.

33. Abel Romero-Corral, Virend K. Somers, Justo Sierra-Johnson, Randal J. Thomas, Kent R. Bailey, Maria L. Collazo-Clavell, Thomas G. Allison, Josef Korinek, John A. Batsis, and Francisco Lopez-Jimenez, "Accuracy of Body Mass Index to Diagnose Obesity in the US Adult Population," *International Journal of Obesity* 32, no. 6 (2008): 959–966.

34. Reubin Andres, "Effect of Obesity on Total Mortality," *International Journal of Obesity* 4, no. 4 (1980): 381–386.

35. These three surveys included 36,859 participants. Data was initially collected from 1971 to 1975 for NHANES I, from 1976 to 1980 for NHANES II, and from 1988 to 1994 for NHANES III. Mortality data was collected through 1992 for NHANES I and II and through 2000 for NHANES III. Katherine M. Flegal, Barry I. Graubard, David F. Williamson, and Mitchell H. Gail, "Excess Deaths Associated with Underweight, Overweight, and Obesity," *JAMA* 293, no. 15 (2005): 1861–1867.

36. Hans T. Waaler, "Height, Weight, and Mortality: The Norwegian Experience," *Acta Medica Scandinavica Supplementum* 679 (1984): 1–56, cited in Paul Campos, *The Obesity Myth: Why America's Obsession with Weight Is Hazardous to Your Health* (New York: Gotham Books, 2004), 10.

37. Dongfeng Gu, Jiang He, Xiufeng Duan, Kristi Reynolds, Xigui Wu, Jing Chen, Guangyong Huang, Chung-Shiuan Chen, and Paul K. Whelton, "Body Weight and Mortality among Men and Women in China," *JAMA* 295, no. 7 (2006): 227–283; Volker Arndt, Dietrich Rothenbacher, Bernd Zschenderlein, Stephan Schuberth, and Hermann Brenner, "Body Mass Index and Premature Mortality in Physically Heavily

Working Men—a Ten-Year Follow-Up of 20,000 Construction Workers," *Journal of Occupational and Environmental Medicine* 49, no. 8 (2007): 913–921; Esa Laara and Paula Rantakallio, "Body Size and Mortality in Women: A 29 Year Follow Up of 12,000 Pregnant Women in Northern Finland," *Journal of Epidemiology and Community Health* 50, no. 4 (1996): 408–414, all cited in Linda Bacon, *Health at Every Size: The Surprising Truth about Your Weight* (Dallas, TX: Benbella Books, 2008), 127.

38. Richard Troiano, Edward A. Frongillo Jr., Jeffrey Sobal, and David Levitsky, "The Relationship between Body Weight and Mortality: A Quantitative Analysis of Combined Information from Existing Studies," *International Journal of Obesity and Related Metabolic Disorders* 20, no. 1 (1996): 63–75.

39. J. Eric Oliver, *Fat Politics: The Real Story behind America's Obesity Epidemic* (New York: Oxford University Press, 2006), 22–23.

40. Katherine M. Flegal, Brian K. Kit, Heather Orpana, and Barry I Graubard, "Association of All-Cause Mortality with Overweight and Obesity Using Standard Body Mass Index Categories: A Systematic Review and Meta-analysis," *JAMA* 309, no. 1 (2013): 71–82.

41. Paul Campos, "Our Absurd Fear of Fat," *New York Times*, January 2, 2013, http://www.nytimes.com/2013/01/03/opinion/our-imaginary-weight-problem.html.

42. David B. Allison, Kevin R. Fontaine, JoAnn E. Manson, June Stevens, and Theodore B. VanItallie, "Annual Deaths Attributable to Obesity in the United States," *JAMA* 282, no. 16 (1999): 1530–1538.

43. Katherine M. Flegal, Barry I Graubard, David F. Williamson, and Mitchell H. Gail., "Excess Deaths Associated with Underweight, Overweight, and Obesity," *JAMA* 293, no. 15 (2005): 1861–1867.

44. Cheri L. Olson, Howard D. Schumaker, and Barbara P. Yawn, "Overweight Women Delay Medical Care," *Archives of Family Medicine* 3, no. 10 (1994): 888–892; Rebecca Puhl and Kelly Brownell, "The Stigma of Obesity: A Review and Update," *Obesity* 17, no. 5 (2009): 941–964.

45. WHO Consultation on Obesity, *Obesity: Preventing and Managing the Global Epidemic*, WHO Technical Report Series 894 (Geneva: World Health Organization, 2000), 46.

46. CDC Division of Cancer Prevention and Control, "Lung Cancer: What Are the Risk Factors?" CDC.gov, November 21, 2013, http://www.cdc.gov/cancer/lung/basic_info/risk_factors.htm.

47. Gary Taubes, "Epidemiology Faces Its Limits," *Science* 14, no. 269 (1995): 164–169.

48. Maria Teresa Guagnano, V. Pace-Palitti, C. Carrabs, D. Merlitti, and S. Sensi, "Weight Fluctuations Could Increase Blood Pressure in Android Obese Women," *Clinical Science* 96, no. 6 (1999): 677–680.

49. In a comparison of adult women in the Mississippi Band of Choctaw, American Samoan women, and African American women in West Alabama who were surveyed in the early to mid-1990s, the Samoans were much more likely to be obese than the other groups but had lower rates of hypertension and diabetes. Similarly, a study involving several ethnic minorities in Xinjiang, China (Kazaks, Uyghurs, Mongolians, and Han immigrants) found that the association between obesity and hypertension in all of the groups was lower than in Western countries and that although the

Mongolians had the highest prevalence of overweight/obesity, they had the lowest prevalence of hypertension. Jim Bindon, William W. Dressler, M. Janice Gilliland, and Douglas E. Crews, "A Cross-Cultural Perspective on Obesity and Health in Three Groups of Women: The Mississippi Choctaw, American Samoans, and African Americans," *Collegium Antropologicum* 31, no. 1 (2007): 47–54; Xiao-Guang Yao, Florian Frommlet, Ling Zhou, Feiya Zu, Hong-Mei Wang, Zhi-Tao Yan, Wen-Li Luo, Jing Hong, Xin-Ling Wang, and Nan-Fang Li, "The Prevalence of Hypertension, Obesity, and Dyslipidemia in Individuals over 30 Years of Age Belonging to Minorities from the Pasture Area of Xinjiang," *BMC Public Health* 10 (February 2010): 91.

50. Ancel Keys, Josef Brozek, Austin Henschel, Olaf Mickelsen, and Henry Longstreet Taylor, *The Biology of Human Starvation*, vol. 1 (Minneapolis: University of Minnesota Press, 1950), cited in Campos, *Obesity Myth*, 21.

51. Bacon, *Health at Every Size*, 132.

52. Elizabeth Barrett-Connor and K. T. Khaw, "Is Hypertension More Benign When Associated with Obesity?" *Circulation* 72 (1985): 53–60; Francois Cambien, Jean M. Chretien, Pierre Ducimetiere, Louis Guize, and Jacques L. Richard, "Is the Relationship between Blood Pressure and Cardiovascular Risk Dependent on Body Mass Index?" *American Journal of Epidemiology* 122 (1985): 434–442; Seth Uretsky, Franz H. Messerli, Sripal Bangalore, Annette Champion, Rhonda M. Cooper-DeHoff, Qian Zhou, and Carl J. Pepine, "Obesity Paradox in Patients with Hypertension and Coronary Artery Disease," *American Journal of Medicine* 120, no. 10 (2007): 863–870.

53. Mercedes R. Carnethon, Peter John D. De Chavez, Mary L. Biggs, Cora E. Lewis, James S. Pankow, Alain G. Bertoni, Sherita H. Golden, et al., "Association of Weight Status with Mortality in Adults with Incident Diabetes," *JAMA* 308, no. 20 (2012): 2085. Carolyn Ross, Robert D. Langer, and Elizabeth Barrett-Connor, "Given Diabetes, Is Fat Better Than Thin?" *Diabetes Care* 20, no. 4 (1997): 650–652.

54. Graham A. Colditz, Walter C. Willett, Andrea Rotnitzky, and JoAnn E. Manson., "Weight Gain as a Risk Factor for Clinical Diabetes Mellitus in Women," *Annals of Internal Medicine* 122, no. 7 (1995): 481–486.

55. Named for the epidemiologist Peter Bennett who developed the theory in the 1980s. Peter Bennett, "More About Obesity and Diabetes," *Diabetologia* (1986): 753–754, cited in Paul Ernsberger and Richard J. Koletsky, "Biomedical Rationale for a Wellness Approach to Obesity: An Alternative to a Focus on Weight Loss," *Journal of Social Issues* 55, no. 2 (1999): 221–260.

56. Ron J. Sigal, Mona El-Hashimy, Blaise C Martin, J. Stuart Soeldner, Andrzej S Krolewski, and James H Warram, "Acute Post-Challenge Hyperinsulinemia Predicts Weight Gain: A Prospective Study," *Diabetes* 46, no. 6 (1997): 1025–1029; Marek Straczkowski, I. Kowalska, A. Stepień, S. Dzienis-Straczkowska, M. Szelachowska, I. Kinalska, A. Krukowska, and M. Konicka, "Insulin Resistance in the First-Degree Relatives of Persons with Type 2 Diabetes," *Medical Science Monitor* 9, no. 5 (2003): 186–190; Kitt Falk Petersen, Sylvie Dufour, Douglas Befroy, Rina Garcia, and Gerald I. Shulman, "Impaired Mitochondrial Activity in the Insulin-Resistant Offspring of Patients with Type 2 Diabetes," *New England Journal of Medicine* 350, no. 7 (2004): 639–641; Kitt Falk Petersen, S. Dufour, D. B. Savage, S. Bilz, G. Solomon, S. Yonemitsu, G. W. Cline, D. Befroy, L. Zemany, B. B. Kahn, X. Papademetris, D. L. Rothman, G. I. Shulman, "The Role of Skeletal Muscle Insulin Resistance in the Pathogenesis of the

Metabolic Syndrome," *Proceedings of the National Academy of Sciences* 104, no. 31 (2007): 12587-12594.

57. Nishi Chaturvedi and John H. Fuller, "Mortality Risk by Body Weight and Weight Change in People with NIDDM: The WHO Study of Vascular Disease in Diabetes," *Diabetes Care* 18, no. 6 (1995): 766-774.

58. Albert Stunkard and Mavis McLaren-Hume, "The Results of Treatment for Obesity: A Review of the Literature and Report of a Series," *Journal of the American Medical Association Archives of Internal Medicine* 103, no. 1 (1959): 79-85.

59. Jane Fritsch, "95% Regain Lost Weight. Or Do They?" *New York Times*, May 25, 1999, http://www.nytimes.com/1999/05/25/health/95-regain-lost-weight-or-do-they.html.

60. Stunkard and McLaren-Hume, "The Results of Treatment for Obesity," 84.

61. Ibid.

62. Robert W. Jeffrey, Leonard H. Epstein, Terence G. Wilson, Adam Drewnowski, Albert J. Stunkard, and Rena R. Wing, "Long-term Maintenance of Weight Loss: Current Status," *Health Psychology* 19, no. 1 Suppl. (2000): 5-16; Wayne C. Miller, "How Effective Are Traditional Dietary and Exercise Interventions for Weight Loss?" *Medicine & Science in Sports & Exercise* 31, no. 8 (1999): 1129-1134; M. G. Perri and P. R. Fuller, "Success and Failure in the Treatment of Obesity: Where Do We Go from Here?" *Medicine, Exercise, Nutrition, and Health* 4 (1995): 255-272; David M. Garner and Susan C. Wooley, "Obesity Treatment: The High Cost of False Hope," *Journal of the American Dietetic Association* 91, no. 10 (1991): 1248-1251; Robert W. Jeffrey, "Behavioral Treatment of Obesity," *Annals of Behavioral Medicine* 9, no. 1 (1987): 20-24; Gerald A. Bennett, "Behavior Therapy for Obesity: A Quantitative Review of the Effects of Selected Treatment Characteristics on Outcome," *Behavior Therapy* 17 (1986): 554-562; K. D. Brownell and T. A. Wadden, "Behavior Therapy for Obesity: Modern Approaches and Better Results," in *The Handbook of Eating Disorders: Physiology, Psychology, and Treatment of Obesity, Anorexia Nervosa, and Bulimia*, ed. K. D. Brownell and J. P. Foreyt (New York: Basic Books, 1986), 180-197; Kelly Brownell, "Obesity: Understanding and Treating a Serious, Prevalent, and Refractory Disorder," *Journal of Consulting and Clinical Psychology* 50 (1982): 820-840; J. P. Foreyt, G. K. Goodrick, and A. M. Gotto, "Limitations of Behavioral Treatment of Obesity: Review and Analysis," *Journal of Behavioral Medicine* 4 (1981): 655-662; G. T. Wilson and K. D. Brownell, "Behavior Therapy for Obesity: An Evaluation of Treatment Outcome," *Advances in Behavior Research Therapy* 3 (1980): 49-86; A. J. Stunkard and S. B. Penick, "Behavior Modification in the Treatment of Obesity: The Problem of Maintaining Weight Loss," *Archives of General Psychiatry* 36 (1979): 801-806; Susan C. Wooley, O. W. Wooley, and S. R. Dyrenforth, "Theoretical, Practical, and Social Issues in Behavioral Treatments of Obesity," *Journal of Applied Behavioral Analysis* 12 (1979): 3-25; J. P. Foreyt, *Behavioral Treatments of Obesity* (New York: Pergamon, 1977); A. J. Stunkard and M. J. Mahoney, "Behavioral Treatment of Eating Disorders," in *The Handbook of Behavior Modification*, ed. H. Leitenberg (Englewood Cliffs, NJ: Prentice-Hall, 1976).

63. See J. P. Montani, A. K. Viecelli, A. Prévot, and A. G. Dulloo, "Weight Cycling during Growth and Beyond as a Risk Factor for Later Cardiovascular Diseases: The 'Repeated Overshoot' Theory," *International Journal of Obesity* 4 (2006): S58-66;

K. Strohacker and B. K. McFarlin, "Influence of Obesity, Physical Inactivity, and Weight Cycling on Chronic Inflammation," *Frontiers in Bioscience* 1, no. 2 (2010): 98–104; Costas A. Anastasiou, Mary Yannakoulia, Vassiliki Pirogianni, Gianna Rapti, Labros S. Sidossis, and Stavros A. Kavouras, "Fitness and Weight Cycling in Relation to Body Fat and Insulin Sensitivity in Normal-Weight Young Women," *Journal of the American Dietetic Association* 110 (2010): 280–284; Juhua Luo, Karen L. Margolis, Hans-Olov Adami, Ana Maria Lopez, Lawrence Lessin, and Weimin Ye, "Body Size, Weight Cycling, and Risk of Renal Cell Carcinoma among Postmenopausal Women: The Women's Health Initiative (United States)," *American Journal of Epidemiology* 166 (2007): 752–759; P. Hamm, R. B. Shekelle, and J. Stamler, "Large Fluctuations in Body Weight during Young Adulthood and Twenty-Five-Year Risk of Coronary Death in Men," *American Journal of Epidemiology* 129 (1989): 312–318; Carlos Iribarren, Dan S. Sharp, Cecil M. Burchfiel, and Helen Petrovitch, "Association of Weight Loss and Weight Fluctuation with Mortality among Japanese American Men," *New England Journal of Medicine* 333 (1995): 686–692; Lauren Lissner, Calle Bengtsson, L. Lapidus, B. Larsson, B. Bengtsson, and Kelly D. Brownell, "Body Weight and Variability in Mortality in the Gothenburg Prospective Studies of Men and Women," in *Obesity in Europe 88: Proceedings of the 1st European Congress on Obesity, June 5–6, 1988, Stockholm, Sweden*, ed. Bjorntorp and Rossner (New Barnet, UK: John Libbey, 1989), 55–60; Kelly D. Brownell and Judith Rodin, "Medical, Metabolic, and Psychological Effects of Weight Cycling," *Archives of Internal Medicine* 154 (1994): 1325–1331.

64. David M. Garner and Susan C. Wooley, "Obesity Treatment: The High Cost of False Hope," *Journal of the American Dietetic Association* 91, no. 10 (1991): 1248–1251.

65. L.W. Craighead, A. J. Stunkard, and R. M. O'Brien, "Behavior Therapy and Pharmacotherapy for Obesity," *Archives of General Psychiatry* 38 (1981): 763–768, quoted in Garner and Wooley, "Obesity Treatment," 1250.

66. S. O. Adams, K. E. Grady, A. K. Lund, C. Mukaida, and C. H. Wolk, "Weight Loss: Long-Term Results in an Ambulatory Setting," *Journal of the American Dietetic Association* 83, no. 3 (1983): 306–310; Patricia M. Dubbert and G. Terrence Wilson, "Goal-Setting and Spouse Involvement in the Treatment of Obesity," *Behavioral Research and Therapy* 22, no. 3 (1984): 227–242; D. S. Kirschenbaum, P. M. Stalonas, Jr., T. R. Zastowny, and A. J. Tomarken, "Behavioral Treatment of Adult Obesity: Attentional Controls and a 2-Year Follow-Up," *Behavioral Research and Therapy* 23, no. 6 (1985): 675–682; J. K. Murphy, B. K. Bruce, and D. A. Williamson, "A Comparison of Measured and Self-Reported Weights in a 4-Year Follow-Up of Spouse Involvement in Obesity Treatment," *Behavior Therapy* 16 (1985): 524–530; B. Rosenthal, G. J. Allen, and C. Winter, "Husband Involvement in the Behavioral Treatment of Overweight Women: Initial Effects and Long-term Follow-up," *International Journal of Obesity* 4 (1980): 165–173, H. Bjorvell and S. Rossner, "Long Term Treatment of Severe Obesity: Four Year Follow Up of Results of Combined Behavioral Modification Programme," *British Medical Journal* 291 (1985): 379–382; L. E. Graham II, C. B. Taylor, M. F. Hovell, and W. Siegel, "Five-year Follow-Up to a Behavioral Weight-loss Program," *Journal of Consulting and Clinical Psychology* 51 (1983): 322–323; H. A. Jordan, A. J. Canavan, and R. A. Steer, "Patterns of Weight Change: The Interval 6 to 10 Years After Initial Weight Loss in a Cognitive-Behavioral Treatment Program," *Psychological Reports* 57 (1985): 195–203; F. M. Kramer, R. W. Jeffery, J. I. Forster, and M. K. Snell, "Long-Term Follow-Up of Behavioral Treatment for Obesity: Patterns of Weight Regain among

Men and Women," *International Journal of Obesity* 13 (1989): 123–136; J. K. Murphy, B. K. Bruce, and D. A. Williamson, "A Comparison of Measured and Self-Reported Weights in a 4 Year Follow-up of Spouse Involvement in Obesity Treatment," *Behavior Therapy* 16(1985): 524–530; P. M. Stalonas, M. G. Perri, and A. B. Kerzner, "Do Behavioral Treatments of Obesity Last? A Five-Year Follow-Up Investigation," *Addictive Behaviors* 9 (1984): 175–183; A. J. Stunkard and S. B. Penick, "Behavior Modification in the Treatment of Obesity: The Problem of Maintaining Weight Loss," *Archives of General Psychiatry* 36 (1979): 801–806, all cited in Garner and Wooley, "Confronting the Failure of Behavioral and Dietary Treatments for Obesity," *Clinical Psychology Review* 11, no. 6 (1991): 729–780.

67. Jeffrey M. Isner, Harold E. Sours, Allen L. Paris, Victor J. Ferrans, and William C. Roberts, "Sudden, Unexpected Death in Avid Dieters Using the Liquid-Protein-Modified-Fast Diet: Observations in 17 Patients and the Role of Prolonged QT Interval," *Circulation* 60, no. 6 (1979): 1401–1412; Harold E. Sours, V. P. Frattali, C. D. Brand, R. A. Feldman, A. L. Forbes, R. C. Swanson, and A. L. Paris, "Sudden Death Associated with Very Low Calorie Weight Reduction Regimens," *American Journal of Clinical Nutrition* 34 (April 1981): 453–461.

68. Robert W. Jeffrey, Leonard H. Epstein, Terence G. Wilson, Adam Drewnowski, Albert J. Stunkard, and Rena R. Wing., "Long-Term Maintenance of Weight-Loss: Current Status," *Health Psychology* 19 (2000): 5–16.

69. Ibid., 14.

70. For example, in a 2007 trial of four popular diets involving over 300 subjects, all groups lost weight in the first six months but then began regaining weight despite still eating significantly fewer calories than before the diets began. At twelve months, the Atkins dieters were eating an average of 289 fewer calories per day, Zone dieters were eating 381 fewer calories, LEARN dieters were eating 271 fewer calories, and Ornish dieters were eating 345 fewer calories. Christopher D. Gardner, Alexandre Kiazand, Sofiya Alhassan, Koowon Kim, Randall S. Stafford, Raymond R. Balise, Helena C. Kraemer, and Abby C. King, "Comparison of the Atkins, Zone, Ornish, and LEARN Diets for Change in Weight and Related Risk Factors among Overweight Premenopausal Women," *JAMA* 297, no. 9 (2007): 969–977. Long-term diet studies have also found that for overweight and obese people who lose weight by dieting and then shift to a maintenance strategy, metabolic and hormonal changes persist long after the diet phase, most of which predispose them to burn fewer calories at rest and experience greater hunger than before they dieted. Priya Sumithran, Luke A. Predergast, Elizabeth Delbridge, Katrina Purcell, Arthur Shulkes, Adamandia Kriketos, and Joseph Proietto, "Long-Term Persistence of Hormonal Adaptations to Weight Loss," *New England Journal of Medicine* 365 (October 2011): 1597–1604.

71. Robert W. Jeffrey and Rena R. Wing, "Long-Term Effects of Interventions for Weight Loss Using Food Provision and Monetary Incentives," *Journal of Consulting in Clinical Psychology* 63, no. 5 (1995): 794.

72. Ibid., 793–796.

73. Gary D. Foster, Holly R. Wyatt, James O. Hill, Brian G. McGluckin, Carrie Brill, Selma Mohammed, Philippe O. Szapary, Daniel J. Rader, Joel S. Edman, and Samuel Klein, "A Randomized Trial of a Low-Carbohydrate Diet for Obesity," *New England Journal of Medicine* 348 (2003): 2082–2090.

74. Gardner et al., "Comparison of Atkins, Zone, Ornish, and LEARN Diets," 975.

75. Iris Shai, Dan Schwarzfuchs, Yaakov Henkin, Danit R. Shahar, Shula Witkow, Ilana Greenberg, Rachel Golan, et al., "Weight Loss with a Low-Carbohydrate, Mediterranean, or Low-Fat Diet," *New England Journal of Medicine* 359 (2008): 229–241.

76. Traci Mann, Janet A. Tomiyama, Erika Westling, Ann-Marie Lew, Barbra Samuels, and Jason Chatman, "Medicare's Search for Effective Obesity Treatments: Diets Are Not the Answer," *American Psychologist* 62, no. 3 (2007): 220–233.

77. Ibid., 230.

78. "NWCR Facts," The National Weight Control Registry, Brown Medical School/ The Miriam Hospital Weight Control and Diabetes Research Center, http://www. nwcr.ws/research/.

79. James O. Hill, Holly Wyatt, Suzanne Phelan, and Rena Wing, "The National Weight Control Registry: Is It Useful in Helping Deal with Our Obesity Epidemic?" *Journal of Nutrition Education Behavior* 37, no. 4 (2005): 169.

80. Ibid.

81. Tara Parker-Pope, "The Fat Trap," *New York Times*, December 28, 2011, http:// www.nytimes.com/2012/01/01/magazine/tara-parker-pope-fat-trap.html.

82. Victoria A. Catenacci, Gary K. Grunwald, Jan P. Ingebrigtsen, John M. Jakicic, Michael D. McDermott, Suzanne Phelan, Rena R. Wing, James O. Hill, and Holly R. Wyatt, "Physical Activity Patterns in the National Weight Control Registry," *Obesity* 16, no. 1 (2008): 153–161.

83. Meghan L. Butryn, Suzanne Phelan, James O. Hill, and Rena R. Wing, "Consistent Self-Monitoring of Weight: A Key Component of Successful Weight Loss Maintenance," *Obesity* 15, no. 12 (2007): 3091–3096.

84. Sio Mei Shick, Rena R. Wing, Mary L. Klem, Maureen T. McGuire, James O. Hill, and Helen Seagle, "Persons Successful at Long-Term Weight Loss and Maintenance Continue to Consume a Low-Energy, Low-Fat Diet," *Journal of the American Dietetic Association* 98, no. 4 (1998): 408–413.

85. Hollie A. Raynor, Robert W. Jeffrey, Suzanne Phelan, James O. Hill, and Rena R. Wing, "Amount of Food Group Variety Consumed in the Diet and Long-Term Weight Loss Maintenance," *Obesity* 13, no. 5 (2005): 883–890; Suzanne Phelan, R. R. Wing, H. A. Raynor, J. Dibello, K. Nedeau, and W. Peng, "Holiday Weight Management by Successful Weight Losers and Normal Weight Individuals," *Journal of Consulting Clinical Psychology* 73, no. 3 (2008): 442–448.

86. Mary Lou Klem, Rena R. Wing, Maureen T. McGuire, Helen M. Seagle, and James O. Hill, "Psychological Symptoms in Individuals Successful at Long-Term Maintenance of Weight Loss," *Health Psychology* 17, no. 4 (1998): 336–345.

87. Catenacci et al., "Physical Activity Patterns in the National Weight Control Registry," 2008.

88. The 45 million statistic was reported by *CBS News* in January 2005, and the 90 million statistic was reported by PR Newswire in August 2010. The difference likely has more to do with the phrasing of the survey question than with a twofold increase in the number of people who were dieting over the span of five years. Christine Lagorio, "Diet Plan Success Tough to Weigh," *CBS News*, January 3, 2005, http://www. cbsnews.com/news/diet-plan-success-tough-to-weigh; PR Newswire, "Cost of Weight Loss in America: Many Americans Would Forgo a Job Promotion to Lose 10 Pounds, Reports Nutrisystem Diet Index," August 12, 2010, http://www.prnewswire.com/

news-releases/cost-of-weight-loss-in-america-many-americans-would-forgo-a-job-promotion-to-lose-10-pounds-reports-nutrisystem-diet-index-100526479.html.

89. PR Newswire, "Cost of Weight Loss in America."

90. Lydia Saad, "To Lose Weight, Americans Rely More on Dieting than Exercise," *Gallup.com*, November 28, 2011, http://www.gallup.com/poll/150986/lose-weight-americans-rely-dieting-exercise.aspx.

91. The study controlled for age, race, smoking, health status, and preexisting disease. The increased risk of mortality was small (1.46–2.70) and cannot be attributed to the weight loss, but the findings do not suggest that weight loss significantly improves health or extends life. D. D. Ingram and M. E. Mussolino, "Weight Loss from Maximum Body Weight and Mortality: The Third National Health and Nutrition Examination Survey Linked Mortality File," *International Journal of Obesity* 34, no. 6 (2010): 1044–1050.

92. Reubin Andres, Denis C. Muller, and John D. Sorkin, "Long-Term Effects of Change in Body Weight on All-Cause Mortality: A Review," *Annals of Internal Medicine* 119, no. 7 (1993): 737–743.

93. Mette K. Simonsen, Y. A. Hundrup, E. B. Obel, M. Grønbaek, and B. L. Heitmann, "Intentional Weight Loss and Mortality among Initially Healthy Men and Women," *Nutrition Reviews* 66, no. 7 (2008): 375–386.

94. In a 2002 survey commissioned by Whole Foods Market, 78 percent of respondents who regularly choose organic products said they taste better and 87 percent say they are of better quality. Whole Foods Market, "One Year after USDA Organic Standards Are Enacted, More Americans Are Consuming Organic Food," *Whole Foods Company News*, October 17, 2002, http://media.wholefoodsmarket.com/news/one-year-after-usda-organic-standards-are-enacted-more-americans-are-consum. In a 2005 survey of current and former CSA members in Iowa, 96 percent of respondents said they became members to get access to "fresh and tasty foods." Corry Bregendahl and Cornelia Butler Flora, "Results from Iowa's Collaborative CSA Member Survey," *North Central Regional Center for Rural Development Research Brief*, October 2006, http://www.soc.iastate.edu/extension/ncrcrd/researchbrief-csamembersurvey-lowres.pdf.

95. *Food, Inc.*, dire. Robert Kenner, Alliance Films, 2008.

96. Office of Campus Sustainability, "University of Michigan—Ann Arbor Sustainability Goal Reporting Guidelines: Goal #4 Purchase 20% of U-M Food in Accordance with Sustainable Food Purchasing Guidelines," *Healthy Environments: Sustainable Food*, November 2012, https://docs.google.com/document/d/1kmofOkOQ8glzCUVFjglBUXseoS2VInEgFWyC18Huo5M/edit?usp=sharing.

97. Ibid.

98. The estimate assumes that twenty mpg is the average for trips with a full load and the diminished post-market load. David Foster, "Tangerines per Gallon," *Chicago Boyz*, July 29, 2007, http://chicagoboyz.net/archives/5111.html.

99. Distance estimate from Christopher L. Weber and H. Scott Matthews, "Food Miles and the Relative Climate Impacts of Food Choices in the United States," *Environmental Science and Technology* 42, no. 10 (2008): 3508–3513. Those fuel efficiency numbers are cited widely, although the EPA figures for Class 8 trucks and data from some new long-haul trucks suggest that they may get close to 10 mpg. Anthony Ingram, "How to Make a Semi Much More Efficient: Cummins 'SuperTruck,'" *Green Car Reports*, March 25, 2013, http://www.greencarreports.com/news/1083123_how-to-make-a-semi-much-more-efficient-cummins-supertruck.

100. At closer to 10 mpg, that would be closer to 393 pounds per gallon, or 196 pounds per gallon counting the return trip. Density is a factor, too. Both trucks will be able to haul more weight in potatoes than lettuce.

101. According to the Association of American Railroads, trains are approximately three to four times as efficient as trucks, moving freight with an average efficiency of 423 ton-miles per gallon. Fuel consumption estimates for shipping containers vary widely but average around 500 ton-miles per gallon. Foster, "Tangerines per Gallon."

102. Christopher L. Weber and H. Scott Matthews, "Food Miles and the Relative Climate Impacts of Food Choices in the United States," *Environmental Science and Technology* 42, no. 10 (2008): 3508–3513.

103. Allison Smith, Paul Watkiss, Geoff Tweddle, Alan McKinnon, Mike Browne, Alistair Hunt, Colin Treleven, et al., *The Validity of Food Miles as an Indicator of Sustainable Development*, ED50254 Issue 7, AEA Technology Environment, July 2005, http://foodsecurecanada.org/sites/default/files/final.pdf.

104. Caroline Saunders, Andrew Barber, and Greg Taylor, *Food Miles: Comparative Energy/Emissions Performance of New Zealand's Agriculture Industry*, Agricultural and Economics Research Unit Report no. 285, July 2006.

105. USDA National Agricultural Statistics Service, *Crop Production: 2013 Summary*, January 10, 2014, http://usda.mannlib.cornell.edu/usda/nass/CropProdSu//2010s/2013/CropProdSu-01-11-2013.pdf.

106. James McWilliams, *Just Food: Where Locavores Get It Wrong and How We Can Truly Eat Responsibly* (New York: Back Bay Books, 2009), 55–58.

107. Consumer shopping trips may be associated with greater energy use and emissions than all other transport, storage, and processing in the supply chain combined. Michael Browne, Christophe Rizet, Stephen Anderson, Julian Allen, and Basile Keita, "Life Cycle Assessment in the Supply Chain: A Review and Case Study," *Transport Reviews* 25, no. 6 (2005): 761–782.

108. Bregendahl and Flora, "Results from Iowa's Collaborative CSA Member Survey," 6.

109. Jacob Leibenluft, "The Overflowing Box of Veggies," *Slate*, October 14, 2008, http://www.slate.com/articles/health_and_science/the_green_lantern/2008/10/the_overflowing_box_of_veggies.html.

110. Claire Hinrichs, "The Practice and Politics of Food System Localization," *Journal of Rural Studies* 19, no. 1 (2003): 33–45, quoted in McWilliams, *Just Food*, 34.

111. Margaret Gray, *Labor and the Locavore: The Making of a Comprehensive Food Ethic* (Berkeley: University of California Press), 2014.

112. Sarah DeWeerdt, "Local Food: The Economics," *World Watch Magazine* 22 (July/August 2009), http://www.worldwatch.org/node/6161.

113. W. S. Pease, R. A. Morello-Frosch, D. S. Albright, A. D. Kyle, and J. C. Robinson, "Preventing Pesticide-Related Illnesses in California Agriculture: Strategies and Priorities," California Policy Seminar, Berkeley, 1993, cited in Julie Guthman, *Agrarian Dreams: The Paradox of Organic Farming in California* (Berkeley: University of California Press, 2004), 124.

114. Jorg Prietzel, Bernhard Mayer, and Allan H. Legge, "Cumulative Impact of 40 Years of Industrial Sulfur Emissions on a Forest Soil in West-Central Alberta," *Environmental Pollution* 131, no. 1 (2004): 129–144.

115. McWilliams, *Just Food*, 68–69.

116. Pierluigi Caboni, Todd B. Sherer, Nanjing Zhang, Georgia Taylor, Hye Me Na, J. Timothy Greenamyre, and John E. Casida, "Rotenone, Deguelin, Their Metabolites, and the Rat Model of Parkinson's Disease," *Chemical Research in Toxicology* 17, no. 11 (2004): 1540–1548.

117. Christie Wilcox, "Mythbusting 101: Organic Farming > Conventional Agriculture," *Scientific American*, July 18, 2011, http://blogs.scientificamerican.com/science-sushi/httpblogsscientificamericancomscience-sushi20110718mythbusting-101-organic-farming-conventional-agriculture.

118. Pamela C. Ronald and Raoul W. Adamchak, *Tomorrow's Table: Organic Farming, Genetics, and the Future of Food* (New York: Oxford University Press, 2008), 98.

119. Joseph Kovach, Harvey Reissig, and Jan Nyrop, "Effect of Botanical Insecticides [on] the New York Apple Pest Complex," Reports from the 1989 IPM Research, Development, and Implementation Projects in Fruit, New York State IPM Program, Cornell University and New York State Department of Agriculture and Markets, IPM Publication #202, 1990, 40–44.

120. Guthman, *Agrarian Dreams*, 22.

121. Crystal Smith-Spangler, Margaret L. Brandeau, Grace E. Hunter, Clay Bavinger, Maren Pearson, Paul J. Eschbach, Vandana Sundaram, et al., "Are Organic Foods Safer or Healthier Than Conventional Alternatives? A Systematic Review," *Annals of Internal Medicine* 157, no. 5 (2012): 348–366.

122. It's possible that some organic farmers are breaking the rules. However, the residue detection techniques are very sensitive, so they may be picking up on traces from spray that drifts from nearby farms or is picked up from equipment or containers used for conventional produce. USDA, *2010–2011 Pilot Study: Pesticide Residue Testing of Organic Produce, USDA National Organic Program and USDA Science and Technology Program*, November 2012, https://www.ams.usda.gov/sites/default/files/media/Pesticide%20Residue%20Testing_Org%20Produce_2010–11PilotStudy.pdf.

123. Bjørn Lomborg, *The Skeptical Environmentalist* (New York: Cambridge University Press, 2001), 233.

124. Average pesticide residue consumption was based on the average dietary intake of fruits and vegetables for 60- to 65-year-old females, which the authors note is higher than the average for other adult age groups. Lois Swirsky Gold, Thomas H. Slone, Bonnie R. Stern, Neela B. Manley, and Bruce N. Ames, "Rodent Carcinogens: Setting Priorities," *Science* 258 (1992): 261–265.

125. Steve Savage, "Pesticides: Probably Less Scary than You Imagine," *Applied Mythology*, September 23, 2012, http://appliedmythology.blogspot.com/2012/09/pesticides-probably-less-scary-than-you.html. Data on 2010 California pesticide use from the State of California, *California Pesticide Information Portal*, http://calpip.cdpr.ca.gov/main.cfm. Pesticide toxicity information is listed in publicly available material safety data sheets. See Rob Toreki, "Where to Find Material Safety Data Sheets on the Internet," *Interactive Learning Paradigms, Incorporated*, 1995–2016, http://www.ilpi.com/msds/#Pesticides. The EPA has four toxicity categories for pesticides. Category I is "highly toxic and severely irritating," category II is "moderately toxic and moderately irritating," category III is "slightly toxic and slightly irritating," and category IV is "practically non-toxic and not an irritant." According to the *PAN Pesticide Database*, Vitamin C would fall into category III and caffeine and aspirin into category II.

S. E. Kegley, B. R. Hill, S. Orme, and A. H. Choi, *PAN Pesticide Database* (Oakland, CA: Pesticide Action Network, 2014).

126. Smith-Spangler et al., "Are Organic Foods Safer?" A response published by the Research Institute of Organic Agriculture points out that organic production systems are highly heterogeneous and that meta-analyses such as the Smith-Spangler study make it difficult to assess how different farming methods might affect nutrient content or pesticide residues. It notes that two studies excluded from the Smith-Spangler because of their failure to specify sample size found a larger difference in phenolic content between organic and conventional produce, attributed to the use of organic fertilizers. Another review study found significantly higher levels of vitamin C (6 percent) and defense-related metabolites (16 percent) in organic produce. Meanwhile, other nutrients such as carotenoids, anthocyanins, and tocopherols that Smith-Spangler found were not significantly different in organic and conventional produce are influenced more by genetic factors and environmental factors, so cultivation method would not be expected to influence those compounds. While it is certainly possible that some organic cultivation methods might yield consistently higher levels of some nutrients, the heterogeneity of organic production systems means that there is no guarantee that produce bearing an organic certification label was grown using those methods. There is also scant evidence that consuming produce with increased nutrient levels in the range of 6 percent higher levels of vitamin C and 16 percent higher levels of defense-related metabolites leads to significant, positive health outcomes. Alberta Velimirov and Thomas Lindenthal, "Opinion on the Publication of the Stanford University Medical School Study: 'Are Organic Foods Safer or Healtheir than Conventional Alternatives? A Systematic Review," *Forschungsinstitut fr biologischen Landbau sterreich/Austria*, October 15, 2012, https://www.fibl.org/fileadmin/documents/en/news/2012/stanford_stellungnahme_lindenthal_englisch_121106.pdf.

127. The current export of guano to wealthy nations where organic produce is in high demand reprises some aspects of the nineteenth-century imperialist scramble for guano. Then, as now, it was valued as a nitrogen-rich fertilizer. In the nineteenth century, the United States actually authorized citizens to forcibly occupy islands where guano was found. Although guano is technically renewable, it is a limited resource, and some people predict that Peru's supplies will be exhausted within the next decade or two. Simon Romero, "Peru Guards Its Guano as Demand Soars," *New York Times*, May 30, 2008, http://www.nytimes.com/2008/05/30/world/americas/30peru.html.

128. McWilliams, *Just Food*, 72–79.

129. Data on the carbon footprint of manure composting from Xiying Hao, Chi Chang, Francis J. Larney, and Greg R. Travis, "Greenhouse Gas Emissions during Cattle Feedlot Manure Composting," *Journal of Environmental Quality* 30 (2001): 376–386. See also Steve Savage, "The Shocking Carbon Footprint of Compost," Applied Mythology, January 9, 2013, http://appliedmythology.blogspot.com/2013/01/the-shocking-carbon-footprint-of-compost.html. Savage suggests that a better use of the manure, ecologically, would be to process it in an anaerobic digester that captures the methane to burn as a form of renewable energy with carbon neutral emissions.

130. Catherine Badgley, Jeremy Moghtader, Eileen Quintero, Emily Zakem, M. Jahi Chappell, Katia Avilés-Vázquez, Andrea Samulon, and Ivette Perfecto, "Organic Agriculture and the Global Food Supply," *Renewable Agriculture and Food Systems* 22, no. 2 (2006): 86–108.

131. David Pimentel, Paul Hepperly, James Hanson, David Douds, and Rita Seidel, "Environmental, Energetic, and Economic Comparisons of Organic and Conventional Farming Systems," *BioScience* 55, no. 7 (2005): 573–582.

132. Comparable systems were those where both systems have the same nonfood rotation length and similar nitrogen inputs. Verena Seufert, Navin Ramankutty, and Jonathan A. Foley, "Comparing the Yields of Organic and Conventional Agriculture," *Nature* 485 (May 10, 2012): 229–234.

133. USDA, *2007 Census of Agriculture, Organic Production Survey 3*, vol. 3, special studies, part 2 ([Washington, DC]: United States Department of Agriculture and National Agricultural Statistics Service, 2008), https://www.agcensus.usda.gov/Publications/2007/Online_Highlights/Organics/ORGANICS.pdf. See also Jayson Lusk, *The Food Police: A Well-Fed Manifesto about the Politics of Your Plate* (New York: Crown Forum, 2013), 93.

134. Online survey of 670 American adults commissioned by Vital Farms and conducted by the consumer research firm Quick Take. Adele Douglass, "New Research Shows Egg Labels Scramble the Minds of Consumers," *Certified Humane*, June 24, 2014, http://certifiedhumane.org/new-research-shows-egg-labels-scramble-minds-consumers.

135. Office of Campus Sustainability, "University of Michigan—Ann Arbor Sustainability Goal Reporting Guidelines: Goal #4 Purchase 20% of U-M Food in Accordance with Sustainable Food Purchasing Guidelines," *Healthy Environments: Sustainable Food*, November 2012, https://docs.google.com/document/d/1kmofOkOQ8glzCUVFjglBUXseoS2VInEgFWyC18Hu05M/edit?usp=sharing.

136. Catherine Price, "Sorting through the Claims of the Boastful Egg," *New York Times*, September 16, 2010, http://www.nytimes.com/2008/09/17/dining/17eggs.html.

137. Anders Kelto, "Farm Fresh? Natural? Eggs Not Always What They're Cracked Up to Be," *NPR*, December 23, 2014, http://www.npr.org/sections/thesalt/2014/12/23/370377902/farm-fresh-natural-eggs-not-always-what-they-re-cracked-up-to-be.

138. Defined by the National Organic Standards Board's Livestock Committee as outdoor space sufficient "to satisfy [the chickens'] natural behavior patterns, provide adequate exercise area, provide preventive health care benefits and answer consumer expectations of organic livestock management." The Cornucopia Institute, *Scrambled Eggs: Separating Factory Farm Egg Production from Authentic Organic Agriculture* (Cornucopia, WI: Cornucopia Institute, 2010), http://www.cornucopia.org/egg-report/scrambledeggs.pdf.

139. The AVMA is a nonprofit association of licensed veterinary practitioners and academics with no clear commercial interest in conventional or alternative egg production. "Welfare Implications of Laying Hen Housing: Literature Review," *AVMA*, March 2010, https://www.avma.org/KB/Resources/LiteratureReviews/Pages/Welfare-Implications-of-Laying-Hen-Housing.aspx.

140. Marion Nestle, "Does the Color of an Egg Yolk Indicate How Nutritious It Is?" *Chowhound*, June 15, 2009, http://www.chowhound.com/food-news/55099/does-the-color-of-an-egg-yolk-indicate-how-nutritious-it-is/.

141. Based on a 2005 analysis of eggs from four heritage-breed pastured flocks and a 2007 analysis of fourteen producers, both commissioned by *Mother Earth News*. Cheryl Long and Tabitha Alterman, "Meet Real Free-Range Eggs," *Mother Earth News*,

October/November 2007, http://www.motherearthnews.com/real-food/free-range-eggs-zmaz07onzgoe.aspx.

142. G. Cherian, T. B. Holsonbake, and M. P. Goeger, "Fatty Acid Composition and Egg Components of Specialty Eggs," *Poultry Science* 81, no. 1 (2002): 30–33.

143. H. D. Karsten, P. H. Patterson, R. Stout, and G. Crews, "Vitamins A, E, and Fatty Acid Composition of the Eggs of Caged Hens and Pastured Hens," *Renewable Agriculture and Food Systems* 25 (2010): 45–54.

144. Joanna Lott, "Pasture-ized Poultry," *Penn State News*, May 1, 2003, http://news.psu.edu/story/140750/2003/05/01/research/pasture-ized-poultry.

145. Cherian et al., "Fatty Acid Composition"; Dan Charles, "What the Rise of Cage-Free Eggs Means for Chickens," *NPR*, June 27, 2013, http://www.npr.org/sections/thesalt/2013/06/27/195639341/what-the-rise-of-cage-free-eggs-means-for-chickens.

146. Nina Shen Rastogi, "The Environmental Impact of Eggs," *Slate*, June 1, 2010, http://www.slate.com/articles/health_and_science/the_green_lantern/2010/06/green_eggs_vs_ham.html.

147. Ibid.

148. Cherian et al., "Fatty Acid Composition."

149. C. E. Realini, S. K. Duckett, G. W. Brito, M. Dalla Rizza, and D. De Mattos, "Effects of Pasture v. Concentrate Feeding with or without Antioxidants on Carcass Characteristics, Fatty Acid Composition, and Quality of Uruguayan Beef," *Meat Science* 66, no. 3 (2004): 567–577, S. K. Duckett, J. P. Neel, R. M. Lewis, J. P. Fontenot, and W. M. Clapham, "Effects of Forage Species or Concentrate Finishing on Animal Performance, Carcass, and Meat Quality," *Journal of Animal Science* 91, no. 3 (2013): 1454–1467.

150. Rajiv Chowdhury, Samantha Warnakula, Setor Kunutsor, Francesca Crowe, Heather A. Ward, Laura Johnson, Oscar H. Franco, et al., "Association of Dietary, Circulating, and Supplement Fatty Acids with Coronary Risk: A Systematic Review and Meta-Analysis," *Annals of Internal Medicine* 160, no. 6 (2014): 398–406. Criticisms of this meta-analysis have focused primarily on an error in one of the studies on omega-3 fats, which was corrected in the online version, and the omission of two studies on omega-6 fats. The precise relationship between saturated fat consumption, heart disease risk, and mortality is the subject of ongoing controversy, and the Chowdury meta-analysis is not the only study to conclude that reducing dietary saturated fat may not reduce the risk of cardiovascular disease or death. A 2009 review concluded that replacing saturated fats with carbohydrates had no benefit, but replacing them with polyunsaturated fats reduced heart disease. A 2015 review concluded that saturated fats are not associated with cardiovascular disease or all-cause mortality. A 2016 analysis of one of the studies that originally led to the belief that saturated fats caused heart disease found that replacing saturated fat with linoleic acid lowered cholesterol but did not translate to a lower risk of death from heart disease or all causes. Kai Kupferschmidt, "Scientists Fix Errors in Controversial Paper About Saturated Fats," *Science Magazine*, March 24, 2014, http://www.sciencemag.org/news/2014/03/scientists-fix-errors-controversial-paper-about-saturated-fats; Marianne U. Jakobsen, Elis J. O'Reilly, Berit L. Heitmann, Mark A. Pereira, Katarina Bälter, Gary E. Fraser, Uri Goldbourt, et al. "Major Types of Dietary Fat and Risk of Coronary Heart Disease: A Pooled Analysis of 11 Cohort Studies," *American Journal of*

Clinical Nutrition 89, no. 5 (2009): 1425–1432; Russell J. de Souza, Andrew Mente, Adriana Maroleanu, Adrian I. Cozma, Vanessa Ha, Teruko Kishibe, Elizabeth Uleryk, "Intake of Saturated and Trans Unsaturated Fatty Acids and Risk of All Cause Mortality, Cardiovascular Disease, and Type 2 Diabetes: Systematic Review and Meta-Analysis of Observational Studies," *BMJ* 351 (August 12, 2015); Christopher E. Ramsden, Daisy Zamora, Sharon Majchrzak-Hong, Keturah R. Faurot, Steven K. Frantz, Robert P. Davis, and John M. Ringel, "Re-Evaluation of the Traditional Diet-Heart Hypothesis: Analysis of Recovered Data from Minnesota Coronary Experiment (1968–1973)" *BMJ* 353 (April 12, 2016).

151. Evangelos C. Rizos, Evangelina E. Ntzani, Eftychia Bika, Michael S. Kostapanos, and Moses S. Elisaf, "Association Between Omega-3 Fatty Acid Supplementation and Risk of Major Cardiovascular Disease Events: A Systematic Review and Meta-analysis," *JAMA* 308, no. 10 (2012): 1024–1033.

152. M. L. Burr, A. M. Fehily, J. F. Gilbert, S. Rogers, R. M. Holliday, P. M. Sweetnam, P. C. Elwood, and N. M. Deadman, "Effects of Changes in Fat, Fish, and Fibre Intakes on Death and Myocardial Reinfarction: Diet and Reinfarction Trial (DART)," *Lancet* 334, no. 8666 (1989): 757–761; M. L. Burr, P. A. L. Ashfield-Watt, F. D. J. Dunstan, A. M. Fehily, P. Breay, T. Ashton, P. C. Zotos, N. A. Haboubi, and P. C. Elwood, "Lack of Benefit of Dietary Advice to Men with Angina: Results of a Controlled Trial," *European Journal of Clinical Nutrition* 57, no. 2 (2003): 193–200.

153. Kenner, *Food, Inc.*; Nina Planck, "Leafy Green Sewage," *New York Times*, September 21, 2006, http://www.nytimes.com/2006/09/21/opinion/21planck.html.

154. Jan M. Sargeant, J. R. Gillespie, R. D. Oberst, R. K. Phebus, D. R. Hyatt, L. K. Bohra, and J. C. Galland, "Results of a Longitudinal Study of the Prevalence of *Escherichia coli* o157:H7 on Cow-Calf Farms," *American Journal of Veterinary Research* 61, no. 11 (2000): 1375–1379; N. Fegan, P. Vanderlinde, G. Higgs, and P. Desmarchelier, "The Prevalence and Concentration of *Escherichia coli* o157 in Faeces of Cattle from Different Production Systems at Slaughter," *Journal of Applied Microbiology* 97, no. 2 (2004): 362–370.

155. L. J. Grauke, S. A. Wynia, H. Q. Sheng, J. W. Yoon, C. J. Williams, C. W. Hunt, and C. J. Hovde, "Acid Resistance of *Escherichia coli* o157:H7 from the Gastrointestinal Tract of Cattle Fed Hay or Grain," *Veterinary Microbiology* 95, no. 3 (2003): 211–225.

156. Aimee N. Hafla, Jennifer W. MacAdam, and Kathy J. Soder, "Sustainability of US Organic Beef and Dairy Production Systems: Soil, Plant and Cattle Interactions," *Sustainability* 5, no. 7 (2013): 3009–3034; Food and Agriculture Organization of the United Nations, "Excessive Grazing and Browsing," Livestock and Environment Toolbox, accessed August 20, 2014, http://www.fao.org/ag/againfo/programmes/en/lead/toolbox/Grazing/Overgrte.htm.

157. G. M. Peters, Hazel V. Rowley, Stephen Wiedemann, Robyn Tucker, Michael D. Short, and Matthias Shulz, "Red Meat Production in Australia: Life Cycle Assessment and Comparison with Overseas Studies," *Environmental Science and Technology* 44, no. 4 (2010): 1327–1332.

158. Charles Kenny, "Got Cheap Milk?" *Foreign Policy*, September 12, 2011, http://foreignpolicy.com/2011/09/12/got-cheap-milk; Will Boisvert, "An Environmentalist on the Lie of Locavorism," *New York Observer*, April 16, 2013, http://observer.com/2013/04/the-lie-of-locavorism; Steve Sexton, "The Inefficiency of Local Food,"

Freakonomics.com, November 14, 2011, http://freakonomics.com/2011/11/14/the-inefficiency-of-local-food. McWilliams, *Just Food*; Lusk, *The Food Police*; Pierre Desrochers and Hiroko Shimizu, *The Locavore's Dilemma: In Praise of the 10,000 Mile Diet* (New York: Public Affairs, 2012).

159. That was the verdict issued at the Carbon Footprint Supply Chain Summit held in London in 2007. McWilliams, *Just Food*, 46.

160. McWilliams, *Just Food*; Guthman, *Agrarian Dreams*; Margaret Gray, *Labor and the Locavore: The Making of a Comprehensive Food Ethic* (Berkeley, CA: University of California Press, 2014).

161. Josee Johnston and Shyon Baumann, *Foodies: Democracy and Distinction in the Gourmet Foodscape* (New York: Routledge, 2010), 69, 98.

162. Edward Said, *Orientalism* (1978; repr., New York: Vintage Books, 1979), 55.

163. "Our Snacks," Frito-Lay, 2014, http://www.fritolay.com/snacks.

164. David Chute, "Fire in the Bowl," *Los Angeles Magazine*, April 2002, 72–75 and 171–172.

165. Rachel Laudan, *Cuisine and Empire: Cooking in World History* (Berkeley: University of California Press, 2013), 28, 164, 244–245, 255.

166. Donna Gabaccia, *We Are What We Eat: Ethnic Food and the Making of Americans* (Cambridge, MA: Harvard University Press, 1998), 150.

167. Andrew Potter, *The Authenticity Hoax* (New York: Harper Perennial, 2010), 133.

CHAPTER 4 — ANYONE CAN COOK

1. *Ratatouille*, dir. Brad Bird, DVD, Walt Disney Pictures, 2007.

2. Christopher Hayes, *Twilight of the Elites: America after Meritocracy* (New York: Crown Publishers, 2012), 51–52.

3. Robert Perrucci and Earl Wysong, *The New Class Society* (Lanham, MD: Rowman & Littlefield, 1999), 186.

4. *Ratatouille*. The repeated references to "heart" are notable in light of the fact that Gusteau's death is attributed to a "broken heart." It may be that even Gusteau, who cannot withstand Ego's criticism, is not as great as Remy.

5. Stacy Finz, "For Its New Film 'Ratatouille,' Pixar Explored Our Obsession with Cuisine," *San Francisco Gate*, June 28, 2007, http://www.sfgate.com/bayarea/article/ BAY-AREA-FLAVORS-FOOD-TALE-For-its-new-film-2583956.php.

6. Pierre Bourdieu, *Distinction: A Social Judgement of Taste*, trans. Richard Nice (New York: Routledge & Kegan Paul, 1984), 196.

7. Based on over $206 million in box office sales earned during its wide release from July 1, 2007, to December 9, 2007. "2007 Domestic Grosses," *Box Office Mojo*, accessed April 20, 2016, http://www.boxofficemojo.com/yearly/chart/?yr=2007. See also "All Time USA Box Office," *Internet Movie Database*, http://www.imdb.com/chart/ boxoffice.

8. "Ratatouille (2007)," *Rotten Tomatoes*, accessed July 23, 2016, https://www. rottentomatoes.com/m/ratatouille/.

9. "Movie Releases by Score: All Time," *Metacritic*, accessed July 3, 2016, http:// www.metacritic.com/browse/movies/score/metascore/all/filtered.

10. A. O. Scott, "Voilà! A Rat for All Seasonings," *New York Times*, June 29, 2007, http://www.nytimes.com/2007/06/29/movies/29rata.html.

11. Justin Chang, "Review: Ratatouille," *Variety*, June 18, 2007, http://variety.com/2007/film/awards/ratatouille-2-1200558501/.

12. Lawrence Topman, "Pixar's Latest Animated Offering Is Rat On," *Charlotte Observer*, June 28, 2007, http://forums.finecooking.com/cookstalk/ipke-kitchen-kvetch/ratatouille-review.

13. Kyle Smith, "'Ratatouille': The Summer's Tastiest Dish," *New York Post*, July 1, 2007, http://nypost.com/2007/07/01/ratatouille-the-summers-tastiest-dish; Scott Foundas, "Ratatouille: Rat Can Cook," *Village Voice*, June 27, 2007, http://www.laweekly.com/film/ratatouille-rat-can-cook-2149508.

14. Of the 609 total reviews, 222 (36 percent) specifically mention the "moral," "ethos," "message," "lesson," or "values" imparted by the film, usually calling it one of the film's assets, if not the primary source of the film's appeal. Only 20 (9 percent) describe the moral as a detraction—e.g., calling it a cliché or unrealistic. Many other reviews that were not counted in the 222 referred to the film's "heart," "soul," or "feel-good story." Only about half of the 222 reviews that specifically mentioned the film's moral also explained what that moral was. Of those, the number describing it as inclusive of anyone or everyone was nearly 40 percent. IMDb users, "Ratatouille," IMDb, http://www.imdb.com/title/tt0382932.

15. IMDb review of *Ratatouille* by Steverino171, http://www.imdb.com/title/tt0382932/reviews?start=361.

16. IMDb review of *Ratatouille* by Neil_fraser, http://www.imdb.com/title/tt0382932/reviews?start=101, viewer review of *Ratatouille*.

17. IMDb reviews of *Ratatouille*: MovieManMA, http://www.imdb.com/title/tt0382932/reviews?start=487; Manny Emmert, http://www.imdb.com/title/tt0382932/reviews?start=50; oneguyrambling, http://www.imdb.com/title/tt0382932/reviews?start=392; sven-goran-lindqvist, http://www.imdb.com/title/tt0382932/reviews?start=434; and snorlax31119, http://www.imdb.com/title/tt0382932/reviews?start=14.

18. IMDb reviews of *Ratatouille*: yadavsonu33, http://www.imdb.com/title/tt0382932/reviews?start=369; and journeymark, http://www.imdb.com/title/tt0382932/reviews-299.

19. IMDb reviews of *Ratatouille*: Peter Gustafsson, http://www.imdb.com/title/tt0382932/reviews?start=339; charmed_halliwells, http://www.imdb.com/title/tt0382932/reviews?start=439; DudeFromDetroit, http://www.imdb.com/title/tt0382932/reviews?start=173; muppet-show, http://www.imdb.com/title/tt0382932/reviews?start=47; Joseph Belanger, http://www.imdb.com/title/tt0382932/reviews?start=125; and Stompga187, http://www.imdb.com/title/tt0382932/reviews?start=213.

20. Melvin Lerner and other social psychologists have developed ways to measure what they call the "just world hypothesis" or "just world fallacy," a tendency of observers to blame victims for their suffering and attribute positive characteristics to people who have received rewards, even when they know that the recipients of either the punishment or reward were selected at random. See Melvin Lerner, *The Belief in a Just World: A Fundamental Delusion* (New York: Plenum, 1980).

21. Richard H. Carmona, "The Obesity Crisis in America," released July 16, 2004, http://www.surgeongeneral.gov/news/testimony/obesity07162003.html.

22. The Nielsen rating/share for adults 18–49 was 4.1/10, meaning that over 4 percent of the adults in America watched *The Biggest Loser* premiere. Steve Rogers,

"NBC's 'The Biggest Loser' Continues Making Ratings Gains," *Reality TV World*, November 23, 2004, http://www.realitytvworld.com/news/nbc-the-biggest-loser-continues-making-ratings-gains-3068.php.

23. Ibid.

24. There are also twenty-five international variations, including *Biggest Loser Asia* filmed in Malaysia, *Suurin pudottaja* (Biggest Loser) in Finland, *O Grande Perdedor* (The Big Loser) in Brazil, *Biggest Loser Jeetega* in India, *XXL* in Latvia, *The Biggest Loser Pinoy Edition* in the Philippines, and *Co masz do stracenia?* (What do you have to lose?) in Poland. Including the original U.S. version, the shows have collectively aired in ninety countries and crowned seventy-six champions. "The Biggest Loser," *Wikipedia*, July 28, 2012, https://en.wikipedia.org/wiki/The_Biggest_Loser.

25. Viewership in season 13 was lower than in past seasons, possibly due to the departure of one of the show's most famous personal trainers, Jillian Michaels, or to new competition in its time slot from *Glee* on Fox and *NCIS* on CBS. It hit another ratings low in 2014. Nonetheless, the show continues to air; as of 2016 it had aired seventeen seasons. "*The Biggest Loser* Hits an All Time Ratings Low," *Herald Sun*, January 28, 2012, http://www.heraldsun.com.au/entertainment/the-biggest-loser-hts-an-all-time-ratings-low/story-e6frf96f-1226255708320.

26. *The Biggest Loser*, season 1, episode 1, aired October 19, 2004, on NBC.

27. I have chosen to refer to the hosts, trainers, and contestants by the names they call each other on the show; in most cases this is their first names. In seasons where multiple contestants shared a name, they usually adopted a nickname (e.g. in season 1, there were two women named Kelly who were referred to throughout the show as "Kelly Mac" and "Kelly Min"). These are also the names that people typically use when they discuss the show in online forums and blogs.

28. *The Biggest Loser*, season 1, episode 1.

29. Ibid.

30. *The Biggest Loser*, season 1, episode 8, aired December 1, 2004, on NBC.

31. *The Biggest Loser*, season 1, episode 1.

32. *The Biggest Loser*, season 3, episode 4, aired October 11, 2006, on NBC.

33. *The Biggest Loser*, season 3, episode 1, aired September 20, 2006, on NBC.

34. *The Biggest Loser*, season 2, episode 2, aired February 5, 2007, on NBC.

35. *The Biggest Loser*, season 4, episode 3, aired September 25, 2007, on NBC.

36. Walter Mischel, Ebbe B. Ebbesen, and Antonette Raskoff Zeiss, "Cognitive and Attentional Mechanisms in Delay of Gratification," *Journal of Personality and Social Psychology* 21, no. 2 (1972): 204–218.

37. *The Biggest Loser*, season 3, episode 1.

38. *The Biggest Loser*, season 1, episode 2, aired November 2, 2004, on NBC.

39. *The Biggest Loser*, season 3, episode 9, aired November 15, 2006, on NBC.

40. *The Biggest Loser*, season 3, episode 7, aired November 1, 2006, on NBC; "The Biggest Loser (Season 4)," *Wikipedia*, March 10, 2011, https://en.wikipedia.org/wiki/The_Biggest_Loser_(season_4).

41. *The Biggest Loser*, season 4, episode 7, aired October 27, 2007, on NBC.

42. Renee, "Be a Biggest Loser," *Fat Fighters*, 2006 accessed November 18, 2008, http://www.fatfighterblogs.com/archives/2006/12/be_a_biggest_lo.php.

43. *The Biggest Loser*, season 3, episode 3, aired October 4, 2006, on NBC.

44. *The Biggest Loser*, season 3, episode 3.

45. *The Biggest Loser*, season 3, episode 12, aired December 13, 2006, on NBC.

46. Comments by goofball and BGSU_Falcon, "IT'S TIME FOR A FEMALE TO WIN!" mynbc.com Discussion Archive for The Biggest Loser: (Season 4), December 19, 2007, accessed March 14, 2011, http://boards.nbc.com/nbc/index. php?showtopic=779652, archived at https://web.archive.org/web/20071224172647/ http://boards.nbc.com/nbc/index.php?showtopic=779652.

47. Comments by matthew_me, gambitsga145, and Ryan_D, ibid.

48. *The Biggest Loser*, season 1, episode 3, aired November 9, 2004, on NBC.

49. littlenicky2, "Who should be in the final four?" mynbc.com Discussion Archive for The Biggest Loser: (Season 4) December 23, 2007 accessed November 18, 2008, http://boards.nbc.com/nbc/index.php?showtopic=779258, archived at https://web. archive.org/web/20071224172555/http://boards.nbc.com/nbc/index.php?showtopic= 779258.

50. *The Biggest Loser*, season 1, episode 10, aired December 21, 2004, on NBC.

51. *The Biggest Loser*, season 3, episode 2, aired September 27, 2006, on NBC.

52. *The Biggest Loser*, season 3, episode 6, aired October 25, 2006, on NBC.

53. Comment by ehaley in response to Sarah Dussault, "Biggest Loser: Do They Eat? Plus Idol Weight Watch," Diet.com, January 30, 2008.

54. lifematters, "Anybody Watching The Biggest Loser This Season?" *Diet.com*, September 22, 2006, accessed October 28, 2008, http://www.diet.com/diet/viewtopic. php?t=5686.

55. Comment by Reneesman in response to ibid.

56. *The Biggest Loser*, season 3, episode 6.

57. *The Biggest Loser*, season 3, episode 10, aired November 29, 2006, on NBC.

58. Ibid.

59. Jennifer Fremlin, "The Weigh-In as National Money Shot," *Flow*, May 7, 2008, http://www.flowjournal.org/2008/05/extreme-biggest-celebrity-fit-loser-makeover-club-the-weigh-in-as-national-money-shot.

60. The season five finale, which aired on April 15, 2008, averaged 3.6 percent of viewing households during the 8 P.M. hour and jumped to 5.1 for the 9 P.M. hour, according to Nielsen's overnight report. Wayne Friedman, "'Biggest Loser' Finale Scores Its Best Rating This Season," *Media Daily News*, April 17, 2008, http://www. mediapost.com/publications/article/80840/biggest-loser-finale-scores-its-best-rating-this.html?edition=19018.

61. Fremlin, "The Weigh-In as National Money Shot."

62. A majority of the contestants have a BMI over 40. As of 2008, only 5.7 percent of the U.S. population had a BMI over 40. Katherine Flegal, Margaret D. Carroll, Cynthia Odgen, and Lester R. Curtin., "Prevalence and Trends in Obesity among US Adults, 1999–2008," *JAMA 303, no. 3* (2010): 235–241.

63. Susan J. Douglas, "We Are What We Watch," *In These Times*, July 1, 2004, http:// inthesetimes.com/article/817/we_are_what_we_watch. Other scholars have echoed this critique, sometimes claiming that the criticism fails to capture the immense variety in the genre but implying that it may be accurate for the "worst" shows. Annette Hill, *Reality TV: Audiences and Popular Factual Television* (New York: Routledge, 2014), 7.

64. Steven Reiss and James Wiltz, "Why People Watch Reality TV," *Media Psychology* 6, no. 4 (2004): 368–378.

65. Lee Siegel, *Not Remotely Controlled: Notes on Television* (New York: Basic Books, 2007), 242–243.

66. *The Biggest Loser*, season 1, episode 11, aired January 4, 2005, on NBC.

67. Brenda Rizzo, "Why the Biggest Loser?" *Brand Liberators*, February 22, 2011, accessed March 17, 2011 http://www.gsw-w.com/blog/2011/02/22/why-the-biggest-loser/.

68. Edward Wyatt, "On 'The Biggest Loser,' Health Can Take a Back Seat," *New York Times*, November 24, 2009, http://www.nytimes.com/2009/11/25/business/media/25loser.html.

69. Weekly sales of both print books and e-books are reported confidentially to the *New York Times* and sales of print titles are statistically weighted to represent all outlets nationwide. "Best Sellers," *New York Times*, March 20, 2011.

70. "Shop the Biggest Loser" *NBCStore.com*, 2012, accessed April 9, 2014, http://www.nbcuniversalstore.com/the-biggest-loser/index.php?v=nbc_the-biggest-loser; Neal Justin, "'Loser' Brand Breaks New Horizons," *Variety TV News*, February 21, 2012, http://variety.com/2012/tv/news/loser-brand-breaks-new-horizons-1118050350.

71. *The Biggest Loser Club*, 2012, accessed April 9, 2014, https://www.biggestloserclub.com.

72. "Reservations," *The Biggest Loser Resort*, 2012, accessed April 09, 2014, http://biggestloserresort.com/rates-and-accommodations.

73. David M. Garner and Susan C. Wooley, "Confronting the Failure of Behavioral and Dietary Treatments for Obesity," *Clinical Psychology Review* 11, no. 6 (1991): 729–780; Robert W. Jeffrey and Rena R. Wing, "Long-Term Effects of Interventions for Weight Loss Using Food Provision and Monetary Incentives," *Journal of Consulting in Clinical Psychology* 63, no. 5 (1995): 794; Robert W. Jeffrey, Leonard H. Epstein, Terence G. Wilson, Adam Drewnowski, Albert J. Stunkard, and Rena R. Wing, , "Long-Term Maintenance of Weight-Loss: Current Status," *Health Psychology* 19 (2000): 5–16; Tracie Mann, Janet A. Tomiyama, Erika Westling, Ann-Marie Lew, Barbra Samuels, and Jason Chatman, "Medicare's Search for Effective Obesity Treatments: Diets Are Not the Answer," *American Psychologist* 62, no. 3 (2007): 220–233.

74. Erin Fothergill, Juen Guo, Lilian Howard, Jennifer C. Kerns, Nicolas D. Knuth, Robert Brychta, Kong Y. Chen, "Persistent Metabolic Adaptation 6 Years After 'The Biggest Loser,'" *Obesity* 24 (2016) http://onlinelibrary.wiley.com/enhanced/doi/10.1002/oby.21538/.

75. Wyatt, "On 'The Biggest Loser,' Health Can Take a Back Seat."

76. "Where Are They Now? Special," episode of *The Biggest Loser*, aired November 24, 2010, on NBC.

77. *Confessions of a Reality Show Loser*, aired January 6, 2009, Discovery Health.

78. *The Biggest Loser*, season 9, episode 19, aired May 25, 2010, on NBC.

79. Julie Rawe, "Fat Chance," *Time*, May 31, 2007 http://content.time.com/time/specials/2007/article/0,28804,1626795_1627112_1626456,00.html.

80. "'Biggest Loser' Contestant: Show 'Hurts' People," *The Early Show*, June 18, 2010, http://www.cbsnews.com/news/biggest-loser-contestant-show-hurts-people.

81. Eric Deggans, "Former 'Biggest Loser' Competitor Kai Hibbard Calls the Show Unhealthy, Misleading," *St. Petersburg Times*, April 4, 2010, http://www.tampabay.

com/features/media/former-biggest-loser-competitor-kai-hibbard-calls-the-show-unhealthy/1084764.

82. "Where Are They Now? Special."

83. Potes, "Thanksgiving Leftovers," *Biggest Loser Recaps*, November 24, 2010, http://www.televisionwithoutpity.com/show/the-biggest-loser-1/where-are-they-now-1.

84. Comment by chocolatine in response to "'Biggest Losers': Where Are They Now?" *Television without Pity*, December 19, 2008, accessed March 15, 2011, http://www.televisionwithoutpity.com/show/the-biggest-loser-1/where-are-they-now.

85. Laura Fraser, *Losing It: False Hopes and Fat Profits in the Diet Industry* (New York: Plume, 1997), 16-50.

86. J. Eric Oliver, *Fat Politics: The Real Story behind America's Obesity Epidemic* (New York: Oxford University Press, 2005), 29.

87. Ibid., 22.

88. Michael Gard and Jan Wright, *The Obesity Epidemic: Science, Morality, and Ideology* (New York: Routledge, 2005), 179.

CHAPTER 5 — JUST MUSTARD

1. It may also have started off as slang among Oxford students to refer to the townsmen, as opposed to the academic gownsmen. For example, a collection of Cambridge student verse includes a description of a local bookseller from 1781: "Snobs call him Nicholson! Plebian name, / Which ne'er would hand a Snobite down to fame / But to posterity he'll go,—perhaps / Since Granta's classic sons have dubbed him Maps." Author unknown, *In Cap and Gown: Three Centuries of Cambridge Wit*, ed. Charles Whibley (London: Kegan Paul, Trench & Co., 1889), 87.

2. Joseph Epstein, *Snobbery: The American Version* (New York: Houghton Mifflin Harcourt, 2003), 13. The *OED* says only that the word is "of obscure origin," although the first definition provided is a colloquialism for shoemaker or cobbler. "snob, n.1," *OED Online*, September 2012, Oxford University Press.

3. Aileen Ribeiro, *Dress and Morality* (London: Berg Publishers, 2003), 12-16.

4. Linda M. Scott, *Fresh Lipstick: Redressing Fashion and Feminism* (New York: Palgrave Macmillan, 2004), 24.

5. "Grey Poupon Original Commercial," YouTube video, https://www.youtube.com/watch?v=G_pGT8Q_tjk.

6. Ibid.

7. Malcolm Gladwell, "The Ketchup Conundrum," *The New Yorker*, September 6, 2004, http://www.newyorker.com/magazine/2004/09/06/the-ketchup-conundrum.

8. Roland Marchand, *Advertising the American Dream: Making Way for Modernity, 1920-1940* (Berkeley: University of California Press, 1985), xvii.

9. Gladwell, "The Ketchup Conundrum."

10. Gillian Teweles, "Brand Management: La Crème de la Mustard," *Madison Avenue* 26, no. 11 (1984): 28. While the term "nouvelle cuisine" has been used to describe many different styles of French cooking dating back to the 1740s, today it is most commonly used to refer to the style popularized in the 1960s by chefs such as Paul Bocuse in a self-conscious rebellion against the older "orthodox" style associated with Georges Auguste Escoffier. The new style is less reliant on rich sauces and complicated techniques, instead emphasizing fresh ingredients prepared so as to

preserve and showcase their original flavor. The term may have been used in that context first by the author Henri Gault to refer to the food Bocuse prepared for the maiden flight of the Concorde airliner in 1969. In the 1980s, when French restaurants became popular again in America, their food was often referred to as "nouvelle cuisine" or "American nouvelle." *France on a Plate*, broadcast November 29, 2008, on BBC Four.

11. Gladwell, "The Ketchup Conundrum."

12. Ibid.

13. Maille was Grey Poupon's primary competition in the Dijon category throughout the 1970s, and as the latter has become more ubiquitous, the former has gained some cachet due to its perceived exclusivity. John Bowen, "Pardon Me. Grey Poupon Is a Niche Brand," *Brandweek*, April, 27, 1998, 14.

14. *Wayne's World*, dir. Penelope Spheeris, DVD, Paramount Pictures, 1992.

15. "Blue Harvest," *Family Guy*, season 6, episode 1, aired September 23, 2007, on FOX.

16. This is one of the most famous lines from *The Brady Bunch*. The list was generated for a week-long special that aired on TVLand December 9–15, 2006. "Baby They're the Greatest," *Florida Times-Union*, December 06, 2006, http://jacksonville.com/tu-online/stories/120606/lif_quotes.shtml#.V5UZOOiAOko.

17. DMX, "We in Here," *Year of the Dog . . . Again*, Columbia Records, 2006.

18. *T.I. vs. T.I.P.* Grand Hustle/Atlantic Records, 2007.

19. "Today's LOLs: All Kittehs," I Can Has Cheezburger?, accessed May 29, 2009, http://icanhas.cheezburger.com/lolcats.

20. For example, Dgllamas writes, "Years ago, my ex b/f was giving my sister & I a ride home. He came upon a stop sign with a distinguished man in the car with a full head of gray hair. My then-b/f actually asked, 'Pardon me, would you have any Grey Poupon?': WE WERE BUSTING A GUT LAAAAAAFFFFFING!!!!!" Also see comments by TregAichi, lilmiss698, and tomack78. Comments in response to "Grey Poupon Original Commercial," YouTube video, https://www.youtube.com/watch?v=G_pGT8Q_tjk.

21. "Grey Poupon 'Son of Rolls' 30 Sec Commercial," YouTube video, https://www.youtube.com/watch?v=NmannAYiwho.

22. Sam Bradley, "ADBANK's Brand Watch," *Brandweek*, May 15, 1995, 44.

23. Pierre Bourdieu, *Distinction: A Social Critique of the Judgment of Taste*, trans. Richard Nice (1979; repr., Cambridge, MA: Harvard University Press, 1984), 247.

24. Some food historians argue that Child's recipes were seen as sophisticated-but-accessible alternatives to the growing range of processed and prepared foods available in the 1950s. Laura Shapiro, *Something from the Oven: Reinventing Dinner in 1950s America* (Bloomington: Indiana University Press, 2004), 74.

25. LeBesco and Naccarato claim that they are not dismissing culinary capital as mere "false consciousness" and that Martha Stewart's fans are "not just cultural dupes." However, the "just" is telling—whatever else the fans of lifestyle programming might be, LeBesco and Naccarato imply that they are also dupes. Culinary capital offers "a means of . . . performing a class identity to which one aspires but that many never actually attain." Kathleen LeBesco and Peter Naccarato, "Julia Child, Martha Stewart, and the Rise of Culinary Capital," in *Edible Ideologies: Representing Food and Meaning*, ed. Kathleen LeBesco and Peter Naccarato (Albany: State University of New York Press, 2007), 235–236.

26. *Sideways*, dir. Alexander Payne, DVD, Fox Searchlight Pictures, 2004.

27. Ibid.

28. Ibid.

29. Ibid.

30. Ibid.

31. Ibid.

32. Ibid.

33. Ibid.

34. Ibid.

35. Ibid.

36. Ibid.

37. Ibid.

38. Ibid.

39. Alexander Payne and Jim Taylor, Sideways Screenplay, The Internet Movie Script Database, http://www.imsdb.com/scripts/Sideways.html.

40. The filmmakers initially tried to get permission to use Chateau Petrus Pomerol instead of Château Cheval Blanc—the former being the most expensive and sought-after Merlot in the world. However, as Christian Moueix, who runs Chateau Petrus, told the *San Francisco Chronicle*, "Quite a few film scripts cross my desk and I vaguely recall 'Sideways' asking for permission to use Petrus. I am afraid that at that time, I found the script unexciting and declined." W. Blake Gray, "Knocked Sideways: Merlot Is Suddenly Uncool—but the Great Ones Still Shine," *San Francisco Chronicle*, February 24, 2005, http://www.sfgate.com/bayarea/article/Knocked-Sideways-Merlot-is-suddenly-uncool-2696453 php.

41. "Absolute Corker," *The Independent*, March 12, 2005, http://www.independent.co.uk/travel/ausandpacific/absolute-corker-528046.html; "Sideways," *Wikipedia*, https://en.wikipedia.org/wiki/Sideways.

42. John Horn, "A Surprise Package: *Juno* Is Catching on in Middle America. $100 Million Looks Possible," *Los Angeles Times*, January 10, 2008, http://articles.latimes.com/2008/jan/10/entertainment/et-word10.

43. "2005 Web Domestic Grosses," *Box Office Mojo*, accessed July 31, 2010, http://www.boxofficemojo.com/yearly/chart/?yr=2005 Web.

44. IMDB reviews of *Sideways* by cowman57-1, http://www.imdb.com/title/tt0375063/reviews?start=225 and bob-2220, http://www.imdb.com/title/tt0375063/reviews?start=375, August 22, 2015.

45. IMDb review of *Sideways* by Sa'ar Vardi, http://www.imdb.com/title/tt0375063/reviews?start=237.

46. IMDb review of *Sideways* by Jaywillingham, http://www.imdb.com/title/tt0375063/reviews?start=299.

47. Review of *Sideways* by PatrickC, *Metacritic*, December 17, 2004, http://www.metacritic.com/movie/sideways/user-reviews?page=3.

48. Richard Kinssies, "On Wine: 'Sideways' Has Intoxicating Effect on Pinot Noir Sales, Some Say," *Seattle Post-Intelligencer*, February 23, 2005, http://www.seattlepi.com/lifestyle/food/article/On-Wine-Sideways-has-intoxicating-effect-on-1167007.php.

49. "Oscar Winner Knocks Sales of Merlot Wine Sideways," *Sunday Times* (London), March 6, 2005, http://www.thesundaytimes.co.uk/sto/news/world_news/article101655.ece.

50. Strawberry Saroyan, "A Night Out with: Virginia Madsen: Days of Wine and Chocolate," *New York Times*, January 16, 2005, http://www.nytimes.com/2005/01/16/fashion/virginia-madsen-days-of-wine-and-chocolate.html.

51. Gray, "Knocked Sideways."

52. Ibid.

53. The difference in magnitude compared to Merlot is at least partially because it represents a far smaller share of the U.S. wine market. Jordan MacKay, Robert Holmes, and Andrea Johnson, *Passion for Pinot: A Journey through America's Pinot Noir Country* (Berkeley, CA: Ten Speed Press, 2009), 13.

54. Stephen Cueller, Dan Karnowsky and Frederick Acosta, "The *Sideways* Effect: A Test for Changes in the Demand for Merlot and Pinot Noir Wines," American Association of Wine Economists Working Papers no. 25, October 2008, http://www.wine-economics.org/workingpapers/AAWE_WP25.pdf.

55. Ibid., 21.

56. Cueller et al. also sought to determine whether the effect was different for different price segments by reexamining the data for wines under $10, $10 to less than $20, and $20 to 40. They hypothesized that the film's effect would have a stronger effect on lower-priced wines. The results for Merlot were mixed, with decreases in every price range. For Pinot Noir, the results were contrary to the hypothesis. The increase in sales growth was smallest in the lowest price segment, and "drastic" for promoted wines costing $20–$40. This suggests that wine consumers at all levels were affected. Ibid., 14–15.

57. Simon Owen, "The *Sideways* Offensive: Will Merlot Sales Ever Recover?" *Bloggasm*, July 3, 2007, http://bloggasm.com/the-sideways-offensive-will-merlot-sales-ever-recover. According to the website of Baker's winery, Dover Canyon, they no longer offer wines made with Merlot grapes. "Our Wine," *Dover Canyon Winery*, http://www.dovercanyon.com/our-wine.html.

58. Greene started her career as a writer for *Cosmopolitan* and the *Ladies' Home Journal*, largely focusing on "sex and the single woman." In 1968, she was hired by *New York Magazine* to review restaurants, although she is best known for writing about sex in the guise of food (or vice versa). David Kamp, *The United States of Arugula: How We Became a Gourmet Nation* (New York: Clarkson Potter, 2006), 134–135.

59. Gael Greene, "What's Nouvelle? La Cuisine Bourgeoisie," *New York Magazine*, June 2, 1980, 33, archived at The Insatiable Critic: Vintage Insatiable, http://www.insatiable-critic.com/Article.aspx?id=1131.

60. Ibid.

61. As recounted by *The Foodie Handbook* coauthor Paul Levy in "What Is a Foodie?" *Word of Mouth Blog*, June 14, 2007, http://www.theguardian.com/lifeandstyle/wordofmouth/2007/jun/14/whatisafoodie.

62. Ibid.

63. Paul Levy and Ann Barr, *The Official Foodie Handbook (Be Modern—Worship Food)* (New York: Timbre Books, 1984), 7, quoted in Josée Johnston and Shyon Baumann, *Foodies: Democracy and Distinction in the Gourmet Foodscape* (New York: Routledge, 2010), 53.

64. Levy, "What Is a Foodie?"

65. Ibid.

66. Comments by gastrotom and Magpiec13 in response to ibid, https://www. theguardian.com/lifeandstyle/wordofmouth/2007/jun/14/whatisafoodie#comments. Using asterisks in place of vowels is a common way to avoid profanity filters; thus, the implication here is that foodie is vulgar.

67. Katharine Shilcutt, "Has the 'Foodie' Backlash Begun?" *Houston Press*, August 9, 2010, http://www.houstonpress.com/restaurants/has-the-foodie-backlash-begun-6431665.

68. "Crème Fraiche," *South Park*, season 14, episode 14, aired November 17, 2010, on Comedy Central.

69. Christopher Borelli, "Foodie Fatigue," *Chicago Tribune*, December 27, 2010, http://articles.chicagotribune.com/2010-12-27/features/ct-live-1227-foodie-backlash-20101227_1_foodies-breakfasts-plate.

70. Carey Polis, "Foodie Backlash Has Come Swiftly, Was Inevitable," *Huffington Post*, October 29, 2012, http://www.huffingtonpost.com/2012/10/29/foodie-backlash_n_2038061.html.

71. Borelli, "Foodie Fatigue."

72. Johnston and Baumann, *Foodies*, 61.

73. Ibid.

74. Ibid, 37.

75. Ibid., 1–2.

76. Johnston and Baumann offer a similar analysis of omnivorousness later in the book, concluding that "because overt snobbery is no longer acceptable, the contemporary cultural consumption strategy must move away from the traditional highbrow model. But because cultural consumption is still related to distinction processes and to individuals' social locations, symbolic boundaries must be maintained. Omnivorousness is the solution to this problem." Ibid., 197–198.

77. Richard Peterson and Roger M. Kern, "Changing Highbrow Taste: From Snob to Omnivore," *American Sociological Review* 61, no. 5 (1996): 900–907.

CHAPTER 6 — FEELING GOOD ABOUT WHERE YOU SHOP

1. Christian Lander, "Full List of Stuff White People Like," *Stuff White People Like*, March 4, 2008, https://stuffwhitepeoplelike.com/full-list-of-stuff-white-people-like. Many of the comments note the apparent elision between whiteness and wealth, describing the list as a whole as more yuppie than white, more about class than race, or as a commenter said of the Whole Foods item, it's only true of "rich white people oh maybe i should say whites with good credit." Comment by nia in response to clander, "#48 Whole Foods and Grocery Co-ops," *Stuff White People Like*, February 3, 2008, https://stuffwhitepeoplelike.com/2008/02/03/48-whole-foods-and-grocery-co-ops.

2. clander, "#48 Whole Foods."

3. Ian Brown, "Author Michael Pollan Explains the War on Food Movement," *Toronto Globe and Mail*, March 18, 2011, http://www.theglobeandmail.com/life/author-michael-pollan-explains-the-war-on-food-movement/article573363.

4. Kim Severson, "Some Good News on Food Prices," *New York Times*, April 2, 2008, http://www.nytimes.com/2008/04/02/dining/02cheap.html.

5. "Burning Question: Is the Organic Food Movement Elitist?" *Plenty Magazine*, April 1, 2007, http://www.plentymag.com/magazine/burning_question_is_the_organi.php.

6. Mark Bittman, "Is Junk Food Really Cheaper?" *New York Times*, September 24, 2011   http://www.nytimes.com/2011/09/25/opinion/sunday/is-junk-food-really-cheaper.html.

7. Eric Schlosser, "Why Being a Foodie Isn't 'Elitist,'" *Washington Post*, April 29, 2011,   https://www.washingtonpost.com/opinions/why-being-a-foodie-isnt-elitist/2011/04/27/AFeWsnFF_story.html.

8. *Food, Inc.*, dir. Robert Kenner, DVD, Alliance Films, 2008.

9. Michael Pollan, "Voting with Your Fork," *New York Times*, May 7, 2006, http://pollan.blogs.nytimes.com/2006/05/07/voting-with-your-fork.

10. "Our Philosophy," SlowFood.com, accessed May 15, 2012, http://www.slowfood.com/about-us/our-philosophy.

11. Nicole Perlroth, "Michael Pollan: The World's 7 Most Powerful Foodies," *Forbes*, November 2, 2011, http://www.forbes.com/sites/nicoleperlroth/2011/11/02/michael-pollan-the-worlds-7-most-powerful-foodies/#3354400261d4.

12. "Take the $5 Challenge," *Slow Food USA*, August 16, 2011.

13. John Birdsall, "Cheap Drama at Slow Food," *Chowhound*, December 14, 2011, http://www.chowhound.com/food-news/101027/slow-food-usa.

14. Ibid.

15. Ibid.

16. Comments by Joe and Noah in response to Jonathan Kauffman, "Slow Food's $5 Challenge Made Alice Waters Cry," *SFWeekly: Dining*, December 16, 2011, http://www.sfweekly.com/foodie/2011/12/16/slow-foods-5-challenge-made-alice-waters-cry.

17. USDA, "Official USDA Food Plans: Cost of Food at Home at Four Levels, U.S. Average, January 2011," January 2011, http://www.cnpp.usda.gov/sites/default/files/usda_food_plans_cost_of_food/CostofFoodJan2011.pdf.

18. "Add Zing to Your $5 Challenge Meal," *Slow Food USA*, December 4, 2011, http://5challenge.tumblr.com/post/13747362923/add-zing-to-your-5-challenge-menu.

19. Americans with annual household incomes lower than $30,000 actually eat fast food significantly less often than people with higher incomes. Households earning $60,000–$70,000 consume the fast food with the highest frequency, nearly twice the rate of households who earn less than $10,000 per year and almost 50 percent more often than households who earn from $10,000 to $30,000. DaeHawn Kim and J. Paul Leigh, "Are Meals at Full-Service and Fast-Food Restaurants 'Normal' or 'Inferior'?" *Population Health Management* 14, no. 6 (2011): 307–315.

20. "Take the $5 Challenge."

21. Kauffman, "Slow Food's $5 Challenge."

22. Birdsall, "Cheap Drama at Slow Food."

23. Gwendolyn Knapp, "A New Chapter of Slow Food NOLA Is Starting Up," *New Orleans Eater*, September 25, 2012, http://nola.eater.com/2012/9/25/6541837/a-new-chapter-of-slow-food-nola-is-starting-up.

24. Many thanks to the reviewer who pointed me to John Coveney's Foucauldian analysis of contemporary nutritional discourse and the obsession with body size. I am making a similar argument about the moralization of "sustainable" eating. John Coveney, *Food, Morals and Meaning: The Pleasure and Anxiety of Eating* (New York: Routledge, 2000).

25. Michel Foucault, *The History of Sexuality*, vol. 2, *The Use of Pleasure*, trans. Robert Hurley (1984; repr., New York: Vintage Books, 1990), 40–41.

26. Ibid., 51–52.

27. Michel Foucault, *The History of Sexuality*, vol. 3, *The Care of the Self*, trans. Robert Hurley (1984; repr., New York: Vintage Books, 1988), 39.

28. Ibid., 140.

29. Francine Prose, *Gluttony* (New York: Oxford University Press, 2006), 13–14.

30. Carol Walker Bynum, *Holy Feast and Holy Fast: The Religious Significance of Food to Medieval Women* (Berkeley: University of California Press, 1987), 2.

31. The chronology he specifies predates the Victorian era by several decades, beginning in the late eighteenth century. Michel Foucault, *The History of Sexuality*, vol. 1, *An Introduction* (1978; repr., New York: Vintage Books, 1990), 116–122.

32. Ibid, 17.

33. Craig Reinarman, "Policing Pleasure: Food, Drugs and the Politics of Ingestion," *Gastronomica* 7 (Summer 2007): 55.

34. Melanie E. Du Puis, "Angels and Vegetables: A Brief History of Food Advice in America," *Gastronomica* 7 (Summer 2007): 36.

35. Benjamin Rush, *Essays, Literary, Moral, and Philosophical* (Philadelphia: Thomas and William Bradford, 1789), 266.

36. Harvey Levenstein, *Paradox of Plenty: A Social History of Eating in Modern America* (1993; repr., Berkeley: University of California Press, 2003), 61. Nobert Elias also claims that the bourgeoisie began to adopt the manners of the nobility in the eighteenth century in *The Civilizing Process: Sociogenetic and Psychogenetic Investigations*, ed. Eric Dunning (1939; repr., Malden, MA: Blackwell Publishers, 2000), 89–99.

37. Joan Jacobs Brumberg, *Fasting Girls: The History of Anorexia Nervosa*, rev. ed. (1988; repr., New York: Vintage Books, 2000), 134.

38. Ibid.

39. John Harvey Kellogg, "Treatment for Self-Abuse and Its Effects," in *Plain Facts for Old and Young* (Burlington, IA: F. Segner & Co., 1891), 302.

40. Ibid., 303.

41. Donna Gabaccia, *We Are What We Eat: Ethnic Food and the Making of Americans* (Cambridge, MA: Harvard University Press, 2000), 124–128.

42. Mary Eberstadt, "Is Food the New Sex? A Curious Reversal in Moralizing," *Policy Review* 153, no. 27 (2009), http://www.hoover.org/research/food-new-sex.

43. Ibid.

44. Ibid.

45. Ibid.

46. Warren Belasco, *Meals to Come: A History of the Future of Food* (Berkeley: University of California Press, 2006), 17.

47. Nick Fiddes, *Meat: A Natural Symbol* (London: Routledge, 1991), 65.

48. Amy Bentley, "Islands of Serenity: Gender, Race, and Ordered Meals during World War II," in *Food in the U.S.A.: A Reader*, ed. Carole M. Counihan (New York: Routledge, 2002), 171–192.

49. Although Americans have become far more supportive of gay marriage in general, there are dramatic and persisting generational differences. In 2015 polls, the support for gay marriage stood at 70 percent among Americans born since 1981 (Millennials), 59 percent among those born between 1965 and 1980 (Generation X), 45 percent of those born between 1946 and 1964 (Baby Boomers), and only 39 percent of

Americans born between 1928 and 1945 (the Silent Generation). Pew Research Center, "Changing Attitudes on Gay Marriage," *Pew Research Center*, July 29, 2015, http://www.pewforum.org/2015/07/29/graphics-slideshow-changing-attitudes-on-gay-marriage.

50. Eberstadt, "Is Food the New Sex?"

51. Foucault, *History of Sexuality*, 1:123.

52. Ibid., 126–127.

53. Ibid., 122.

54. Ibid.

55. J. Eric Oliver, *Fat Politics: The Real Story behind America's Obesity Epidemic* (New York: Oxford University Press, 2006), 36; Michael Gard, *The End of the Obesity Epidemic* (New York: Routledge, 2011), 1.

56. Kim Severson, "Los Angeles Stages a Fast Food Intervention," *New York Times*, August 12, 2008, http://www.nytimes.com/2008/08/13/dining/13calo.html.

57. Erin Durkin, "Vendors See Mixed Results after City's Green Cart Push to Sell Fruit, Veggies in 'Deserts,'" *New York Daily News*, April 27, 2010, http://www.nydailynews.com/new-york/brooklyn/vendors-mixed-results-city-green-cart-push-sell-fruit-veggies-deserts-article-1.167215.

58. Foucault, *History of Sexuality*, 1:127.

59. Roland Sturm and Aiko Hattori, "Diet and Obesity in Los Angeles County 2007–2012: Is There a Measurable Effect of the 2008 'Fast-Food Ban'?" *RAND Social Science and Medicine* 133 (May 2015): 205–211.

60. Durkin, "Vendors See Mixed Results."

61. USDA Economic Research Service, "Access to Affordable and Nutritious Food: Measuring and Understanding Food Deserts and Their Consequences," http://www.ers.usda.gov/media/242675/ap036_1_.pdf.

62. Helen Lee, "The Role of Local Food Availability in Explaining Obesity Risk among Young School-Age Children," *Social Science & Medicine* 74 (April 2012): 1193–1203; An Ruopeng and Roland Sturm, "School and Residential Neighborhood Food Environment and Diet Among California Youth," *American Journal of Preventative Medicine* 42 (February 2012): 129–135.

63. Gina Kolata, "Studies Question the Pairing of Food Deserts and Obesity," *New York Times*, April 17, 2012, http://www.nytimes.com/2012/04/18/health/research/pairing-of-food-deserts-and-obesity-challenged-in-studies.html.

64. Prose, *Gluttony*, 14.

65. Michael Pollan, "You Are What You Grow," *New York Times*, April 22, 2007, http://www.nytimes.com/2007/04/22/magazine/22wwlnlede.t.html.

66. Ibid.

67. Severson, "Some Good News on Food Prices."

68. James McWilliams, "Should We Really Pay $4 for a Peach?" *The Atlantic*, September 7, 2010, http://www.theatlantic.com/health/archive/2010/09/should-we-really-pay-4-for-a-peach/62503.

69. Fred Kuchler and Hayden Stewart, "Price Trends Are Similar for Fruits, Vegetables, and Snack Foods," Economic Research Report no. 55, March 2008, http://www.ers.usda.gov/media/224301/err55.pdf.

70. McWilliams, "Should We Really Pay $4 for a Peach?"

71. DaeHwan and Leigh, "Are Meals at Full-Service and Fast-Food Restaurants 'Normal' or 'Inferior'?"

72. Andrew Dugan, "Fast Food Still Major Part of U.S. Diet," Gallup, August 6, 2013, http://www.gallup.com/poll/163868/fast-food-major-part-diet.aspx.

73. Sundeep Vikraman, Cheryl D. Fryar, and Cynthia L. Ogden, "Caloric Intake from Fast Food among Children and Adolescents in the United States, 2011–2012," *Centers for Disease Control and Prevention*, September 2015, http://www.cdc.gov/nchs/products/databriefs/db213.htm.

74. Share Our Strength and APCO Insight, *It's Dinnertime: A Report on Low-Income Families' Efforts to Plan, Shop for, and Cook Healthy Meals*, January 2012, https://www.nokidhungry.org/images/cm-study/report-full.pdf.

75. Ben Worthen, "A Dozen Eggs for $8? Michael Pollan Explains the Math of Buying Local," *Wall Street Journal*, August 5, 2010, http://www.wsj.com/articles/SB10001424052748704271804575405521469248574.

76. Michael Pollan, *The Omnivore's Dilemma: A Natural History of Four Meals* (New York: Penguin Books, 2006), 11.

77. Worthen, "A Dozen Eggs for $8?"; and Pollan, *The Omnivore's Dilemma*, 11.

78. Stephanie Ogburn, "James McWilliams' Over-Hyped and Undercooked Anti-Locavore Polemic," *Grist*, September 8, 2009, http://grist.org/article/2009-09-08-mcwilliams-locavore-polemic/.

79. Kelly Trueman, "Inflammatory New Book Attacking Local Food Movement Has One Grain of Truth Buried under Heaps of Manure," *AlterNet*, August 26, 2009, http://www.alternet.org/story/142202/inflammatory_new_book_attacking_local_food_movement_has_one_grain_of_truth_buried_under_heaps_of_manure.

80. Rebekah Denn, "Just Food: A Challenge to Current Ideas about Responsible Eating," *Christian Science Monitor*, September 2, 2009, http://www.csmonitor.com/Books/Book-Reviews/2009/0902/just-food.

81. Trueman, "Inflammatory New Book Attacking Local Food Movement Has One Grain of Truth Buried under Heaps of Manure"; Ogburn, "James McWilliams' Over-Hyped and Undercooked Anti-Locavore Polemic."

82. Denn, "Just Food."

83. Jennifer Bleyer, "Hipsters on Food Stamps," *Salon.com*, March 15, 2010, http://www.salon.com/2010/03/16/hipsters_food_stamps_pinched.

84. Ibid.

85. Ibid.

86. Comment by KathyI in response to ibid., http://www.salon.com/2010/03/16/hipsters_food_stamps_pinched/#postID=2027113&page=16&comment=1501861.

87. Comment by pjamma in response to ibid, http://www.salon.com/2010/03/16/hipsters_food_stamps_pinched/#postID=2027113&page=15&comment=1501800.

88. "Letters to the Editor, Re: Hipsters on Food Stamps," *Salon.com*, March 18, 2010 accessed April 08. 2010, http://letters.salon.com/mwt/pinched/2010/03/15/hipsters_food_stamps_pinched/view/?show=all. The letters in response to the article are no longer available, but some quotes (including this one) were reproduced at "Food Stamp Foodies," *EconomyBeat.org*, March 25, 2010, http://economybeat.org/consumers/food-stamp-foodies/index.html.

89. Martin Gilens, *Why Americans Hate Welfare: Race, Media, and the Politics of Antipoverty Policy* (Chicago: University of Chicago Press, 2000), 104.

90. K. K. Skinner, A. G. Anderson, and M. Anderson, eds., *Reagan's Path to Victory: The Shaping of Ronald Reagan's Vision: Selected Writings* (New York: Free Press, 2004), 76.

91. Bleyer, "Hipsters on Food Stamps."

92. The phrase a "chicken in every pot" is widely attributed to President Herbert Hoover, apparently because it was invoked in a 1928 Republican National Committee ad that claimed that Hoover had put the "proverbial 'chicken in every pot.' And a car in the backyard, to boot." The implication was that a vote for Hoover was a vote for continued prosperity. Paul Dickson, *Words from the White House: Words and Phrases Coined or Popularized by America's Presidents* (New York: Walker Publishing Company, Inc., 2013), 43.

93. Bleyer, "Hipsters on Food Stamps."

94. Ibid.

95. Comment by tweeders1 in response to ibid.http://www.salon.com/2010/03/16/hipsters_food_stamps_pinched/#postID=2027113&page=18&comment=1502059.

96. Comment by Aburkett in response to ibid., http://www.salon.com/2010/03/16/hipsters_food_stamps_pinched/#postID=2027113&page=1&comment=1500475.

97. "Perhaps a solution that could make more people happy would be requiring them to take more traditional recipients shopping, and to teach them to use ingredients with which they may not be familiar to cook simple, flavorful, nutritious meals. They might even see the benefit of public service." Comment by SalliganeG in response to ibid., http://www.salon.com/2010/03/16/hipsters_food_stamps_pinched/#postID=2027113&page=10&comment=1500645.

98. Comment by Soliel in response to ibid, http://www.salon.com/2010/03/16/hipsters_food_stamps_pinched/#postID=2027113&page=1&comment=1500474.

99. This suggestion might be difficult to implement, but it isn't actually novel. As Donna Gabaccia notes, at the same time that home economists were trying to Americanize new immigrants by teaching them to eat corn, often prepared in simple dishes that they portrayed as frugal and nutritious, such as Indian corn pudding, the federal Indian Bureau excluded corn from rations, seemingly because the Indians wanted corn and the government thought they could assimilate them with rations of wheat. Gabaccia, *We Are What We Eat*, 130.

100. Foucault, *The History of Sexuality*, 1:5–10.

CONCLUSION

1. Adelle Davis, *Let's Get Well* (New York: Harcourt, Brace & World, 1965), 13.

2. Dan Hurley, *Natural Causes: Death, Lies, and Politics in America's Vitamin and Herbal Supplement Industry* (New York: Random House, 2006), 43.

3. Ibid., 46.

4. Michael Maniates, "Individualization: Plant a Tree, Buy a Bike, Save the World?" *Global Environmental Politics* 1, no. 3 (2001): 31–52; Andrew Szasz, *Shopping Our Way to Safety: How We Changed from Protecting the Environment to Protecting Ourselves* (Minneapolis: University of Minnesota Press, 2007).

Index

About the Author

S. MARGOT FINN has a Ph.D. in American Culture from the University of Michigan, where she is currently a lecturer teaching classes on food, obesity, and the liberal arts. Her previous publications include contributions to *Food Fights: How the Past Matters in Contemporary Food Debates*, edited by Chad Luddington and Matthew Booker (forthcoming); *The Encyclopedia of Food and Agriculture Ethics*, edited by Paul B. Thompson and David Kaplan (2015); *Food Issues: An Encyclopedia*, edited by Ken Albala (2014); *Routledge International Handbook of Food Studies*, edited by Ken Albala (2012); and *A Foucault for the 21st Century*, edited by Sam Binkley and Jorge Capetillo (2009).